Manual of Water Supply Practices

M73

Groundwater Management

American Water Works Association

Manual of Water Supply Practices—M73, First Edition

Groundwater Management

If you find errors in this manual, please email books@awwa.org. Possible errata will be posted at www.awwa.org/M73

Senior Managing Editor/Project Manager: Melissa Valentine
Technical Editor: Dianne Beirne
Cover Design/Technical Illustrations: Michael Labruyere
Production: Senior Specialist - Manuals: Willadee Hitchcock

Library of Congress Cataloging-in-Publication Data

Names: Bloetscher, Frederick, author. | Roberts, Susan V., author. | American Water Works Association, sponsor.
Title: M73 Groundwater Management / by Frederick Bloetscher and Susan Roberts
Other titles: Ground water management | AWWA manual; M73.
Description: First edition. | Denver : American Water Works Association, [2021] | Series: Manual of water supply practice ; M73 | Includes bibliographical references and index. | Summary: "Manual M73 Groundwater Management is the culmination of efforts by members of the Groundwater Resource Committee to identify and demonstrate the value of using computer modeling tools to help utilities better manage their groundwater resources in light of greater competition for water and dwindling supplies in many parts of the world. I"—Provided by publisher.
Identifiers: LCCN 2021016570 | ISBN 9781647170400 (paperback) | ISBN 9781613005781 (Adobe PDF)
Subjects: LCSH: Wells. | Groundwater. | Water-supply--Management.
Classification: LCC TD403 .B57 2021 | DDC 628.1/14--dc23
LC record available at https://lccn.loc.gov/2021016570

Printed in the United States of America

ISBN 978-1-64717-040-0 e-ISBN-13 978-1-61300-578-1
https://doi.org/10.12999/AWWA.M73ed1

American Water Works Association

American Water Works Association
6666 West Quincy Avenue
Denver, CO 80235-3098
awwa.org

Contents

Figures

Tables

This page intentionally blank.

Preface

Manual M73, Groundwater Management, is the culmination of efforts by members of AWWA's Groundwater Resource Committee to identify and demonstrate the value of using computer modeling tools to help utilities better manage their groundwater resources in light of greater competition for water and dwindling supplies in many parts of the world. In North America, we see acute issues developing in the Plains states, Rocky Mountains, desert Southwest, and even along parts of the Eastern Seaboard. In many of these cases, deep and often confined aquifers are being tapped for irrigation and potable supplies, but there is little hope these aquifers can recover their supplies in our lifetime. That creates the potential for serious economic and social consequences later in the 21st century for communities whose wells may run dry and have no backup options. To this end, this manual includes a discussion on the sustainability of groundwater supplies; overdrafting; and the need for groundwater protection, planning, and evaluation efforts that have evolved in the past 10 years. While groundwater management was incorporated to a small degree in the 2014 version of M21, Groundwater, the committee felt the topic has developed considerably over the past 10 years and that a more detailed and focused manual for groundwater management was timely. M21 will now focus primarily on operations.

M73 provides the reader with a general understanding of the principles involved with groundwater, its movement and character, and the subsequent impact these characteristics have on the development of groundwater resources systems and their long-term sustainability for water utilities. This manual should be considered a companion to M21. There are more detailed books available in the AWWA catalog on well drilling procedures, well maintenance, groundwater treatment, and injection programs that are tangential to groundwater management but are relevant to utilities that rely on groundwater for supplies. We encourage interested readers to pursue additional information, some of which is by the authors of this manual. There are also highly technical texts that deal with the mathematics of groundwater, but the math is generally beyond the scope of this manual (see Bloetscher et al. 2005, for example, for these details).

We have specifically not provided much detail on specific groundwater models since they change and anything included would be outdated before this manual was published. Most of the groundwater models are based on the US Geological Survey (USGS) model code, MODFLOW. Added packages that have been developed to complement MODFLOW include SEAWAT, MT3D, MODPATH, and others. Many commercial software providers have taken these base models and incorporated sophisticated pre- and post-processing capabilities to make the models easier to interpret for professionals and policymakers alike. Examples of the results are included in Chapter 7. Modeling has evolved to the point where many water-use projects include a modeling exercise.

Chapter 1 includes a case study of what can go wrong, an overview of groundwater for those not familiar with the topic, and what will be accomplished in this manual. Chapter 2 is an overview of the hydrologic cycle as it relates to the occurrence of groundwater, as well as a discussion of sustainability of groundwater supplies in light of regional competition for water resources and climate change. Chapter 3 starts with a discussion of issues associated with water quality of groundwater and potential impacts from aquifer declines; water quality and contaminant transport resulting from organic, inorganic, and bacteriological pollution; and concepts for addressing source water protection and cleanup. Chapter 4 outlines regional groundwater topics on groundwater recharge, water budgets, characterizing the aquifer, aquifer yield, and some discussion on the regulatory environment. Chapter 5 outlines the monitoring to support management of regional groundwater systems. This includes identifying data and the completeness of same in anticipation of further analysis.

The discussion of using the data from Chapters 1 to 5 to construct a model of the groundwater system begins in earnest in Chapter 6. An outline of common modeling software is included as part of the discussion. The chapter is designed to offer enough detail about how the models are constructed and operate without getting into details of actual software that will inevitably change and therefore be of limited value in a few years. To show the power of these models, Chapter 7 outlines a series of applications of groundwater models. Note: Many of these are fairly complex and may be beyond the needs of small utilities, but as the cost of software drops, the ability to create dynamic system models even for small communities increases. This gets beyond the analytical calculations currently used, which have severe limitations when dealing with confined aquifers. The intent of this chapter is to provide a series of modeling case studies from places in North America, using different aquifers and modeling for different purposes. Recharge with reclaimed water (indirect potable reuse potential) along with risk is included in the last case study.

Chapter 8 summarizes the current status of groundwater management and suggests pathways to improve our understanding and policy decision-making with respect to this often-limited resource. This manual should help managers and engineers gain enough background on groundwater management to improve their decision-making, and determine when models should be used and the complexity that should be included. The manual should help these professionals answer many of their questions about complex aquifer systems and improve their ability to manage concerns. The Groundwater Resource Committee is hopeful that this manual will meet the industry needs of the new millennium.

Susan Roberts, Ph.D., P.G.
Chair, AWWA Groundwater Resource Committee

Frederick Bloetscher, Ph.D., P.E.
Manual Editor, AWWA Groundwater Resource Committee

Acknowledgments

This manual was developed by the AWWA Groundwater Resource Committee. The membership at the time it approved this manual was as follows:

Susan Roberts (Chair), San Antonio, Tex.
Frederick Bloetscher (Past Chair, incoming Chair), Florida Atlantic University, Boca Raton, Fla.
Angela Bolton, Anchorage Water and Wastewater Utility, Alaska
Steven Colabufo, Suffolk County Water Authority, N.Y.
Leslie Dumas, RMC Water and Environment, Los Angeles, Calif.
Sandra Eberts, US Geological Survey, Columbus, Ohio
David F. Edson, Jensen Beach, Fla.
Dena Egenhoff, City of Cheyenne, Wyo.
Matthew Gamache, CDM Smith, Boston, Mass.
Alex Gerling, Lakewood, Colo.
David Jordan, Albuquerque, N.M.
Mitch Kannenberg, WSP USA Inc., Sioux Falls, S.Dak.
Sharon C. Long, retired, Madison, Wis.
Albert Muniz, Hazen & Sawyer, Boca Raton, Fla.
Willadee Hitchcock (Senior Specialist – Manuals), AWWA, Denver, Colo.

Definitions

1. *Annulus/annular space.* The space between the borehole wall and the well casing, or the space between a casing pipe and a liner pipe.
2. *Aquifer.* A rock formation that will yield water in a usable quantity to a well or spring.
3. *Aquifer mining.* The occurrence whereby the aquifer levels are drawn down with time at a rate faster than recharge.
4. *Artesian aquifer.* Confined aquifer where the water level rises above the top of the aquifer.
5. *Artificial recharge.* Occurs where a well, series of wells, or large basin is created for the purpose of putting water back into a formation. This is different from aquifer storage and recovery, where injection and withdrawal from the same well is planned as a cyclical process.
6. *Bridging.* The development of gaps or obstructions in either grout or filter pack materials during placement or development.
7. *Capillary fringe.* The subsurface layer in which groundwater seeps up from a water table because of tension saturation.
8. *Casing (well).* The vertical pipe that is inserted after drilling, often cemented in place, to give the well structural integrity while sealing off portions of the aquifer that are not part of the production zone. There may be multiple casings in a well.
9. *Conductor casing.* Outer casing used for stability or to seal off a formation to prevent formation collapse or vertical cross-contamination within the well.
10. *Confined aquifer.* An aquifer that is not open to the surface because of some overlying formation.
11. *Confining layer.* Formations that water cannot flow through easily.
12. *Conjunctive use.* Relates to the combined use of groundwater and surface water. The idea of this management approach is to use surface water when the water table is high and change to groundwater when the water table is low, and vice versa. Other sources (reclaimed water) can be incorporated as well.
13. *Decay.* A chemical concentration decrease over time in groundwater, often because of microbial action when related to organics and nutrients.
14. *Dispersion.* The spreading of a contaminant plume from highly concentrated areas to less-concentrated areas.
15. *Filter pack.* Sand, gravel, or glass beads that are uniform, clean, and well-rounded that are placed in the annulus of the well between the borehole wall and the well intake to prevent formation material from entering through the well intake and to stabilize the formation.
16. *Finite difference models.* These computer models convert ordinary linear differential equations or nonlinear partial differential equations into a system of equations that can be solved by matrix algebra techniques, which makes finding the solution well suited to modern computers. The finite difference model is set up using a grid of rectangular grids.

17. *Finite element models.* These computer models convert partial differential equations into a system of equations that can be solved simultaneously in a nonrectangular grid (labeled "elements").
18. *Flowing artesian well.* An artesian well in which the water level stands above the land surface.
19. *Grout.* A fluid mixture of neat cement and water, possibly with various additives or bentonite, of a consistency that can be forced through a pipe and placed in the annular space between the borehole and casing to form a seal.
20. *Head.* Pressure in a formation and/or elevation of water in the formation.
21. *Hydraulic conductivity.* The ability of water to flow through the porous media; referred to as hydraulic conductivity (K).
22. *Permeability.* A measure of the ease with which water can move through a porous rock.
23. *Pressure grouting/sealing.* A process by which grout is confined within the borehole or casings by the use of plugs or packers and by which sufficient pressure is applied to force the grout slurry into and within the annular space or aquifer zone to be grouted.
24. *Sanitary seal.* The top covering of a well, designed to keep insects, surface runoff, small animals, dirt, and debris from entering the well and contaminating the aquifer.
25. *Screen/well intake.* A screening device or length of pipe that is used to keep materials other than the aquifer fluids (groundwater) from entering the well.
26. *Slot size.* The width of the slots machined into a slotted well casing (screen) that allows aquifer fluids (groundwater) into the well.
27. *Sorption.* A physical and chemical process by which one substance becomes attached to another.
28. *Sublimation.* A chemical process where a solid turns into a gas without going through a liquid stage.
29. *Transmissivity.* The hydraulic conductivity multiplied by the aquifer thickness.
30. *Unconfined aquifer.* Shallow formations that have no overlying barrier to direct recharge from precipitation.
31. *Vadose zone.* Includes surface soils, the unsaturated material below the soil, and the capillary fringe (area where water can rise), but is not part of any aquifer system.
32. *Well liner.* Vertical pipe that is inserted inside the casing pipe, connected to the production zone.

This page intentionally blank.

Chapter 1

Introduction

COASTAL CAROLINAS – WHAT CAN GO WRONG?

The eastern Carolinas have long used groundwater as their source water because the aquifer was deep (250–400 ft), confined, stable with respect to water quality, and required little treatment. The communities from Greenville, N.C., through Kinston and Jacksonville, N.C., Myrtle Beach, S.C., and as far south as the communities around Mount Pleasant and Charleston, S.C., withdrew water from formations labeled the Middendorf and Black Creek aquifers. The aquifers were highly productive (calibrated transmissivity values of 400–11,000 ft^2/d for the Black Creek aquifer and 300–30,000 ft^2/d for the Middendorf aquifer), so water was easy to withdraw. In addition, the water required no treatment so communities bragged about the quality of their water that needed no treatment. Richlands, N.C., claimed to be the "Town of Perfect Water."

The Black Creek aquifer is composed of the Cretaceous Black Creek and Middendorf formations (Winner and Coble 1996). The Black Creek aquifer extends across most of the region. It consists of very fine to fine "salt and pepper" sands (NCDENR 2008). The thickness of Black Creek aquifer ranges up to 442 ft. In North Carolina, the Surficial, Castle Hayne, Yorktown, and Peedee aquifers overlie the Black Creek aquifer, and the Upper Cape Fear aquifer underlies the Black Creek (Lautier 2001). The potentiometric or hydraulic head levels of the Black Creek aquifer have been recorded since 1917 (Cooke 1936), and potentiometric maps have been published by Hockensmith (2008, 2003a, 2003b, and 1997), Stringfield and Campbell (1993), and Aucott and Speiran (1985a and 1985b). While the Black Creek aquifer is a source of high-quality drinking water, the presence of saltwater prevents the aquifer from being used for water supply immediately along the coast.

As the population of the eastern Carolinas began to grow in the 1960s, more industry and more people drilled wells into the formation, and water use began to increase rapidly. Myrtle Beach, S.C., became a tourist mecca. Growing Jacksonville, N.C., home of Camp Lejeune, and surrounding Onslow County expanded greatly in the 1980s, expanding their wellfields while deepening their wells. For example, Jacksonville lowered pumps at least 10 ft/yr over 14 years from the late 1970s through the 1980s. Kinston, N.C., was home to major textile and lumber industries that required large volumes of water that also required continual lowering of pumps more than 200 ft. The lowering of pumps began slowly, as all of these communities began to see decreases in water levels starting in the 1970s and accelerating in the 1980s. Ultimately, the aquifer has declined as much as 200 ft from overpumping and depleting storage in North Carolina (North Carolina Department of Environmental Resources 2001; see Figure 1-1).

Similar declines were noted throughout the larger communities in eastern South Carolina as well. Aucott and Speiran (1985b) compared estimates of the predevelopment surface with November 1982 water levels and determined that Black Creek aquifer water levels had declined in Horry and Georgetown counties. Stringfield and Campbell (1993) published 1989 water levels and observed that the water levels in Georgetown, Horry, northern Marion, and northeastern Williamsburg counties had declined since 1982. Data gathered in November 1995 (Hockensmith 1997) and November 2001 (Hockensmith 2003a) showed additional declines influenced by large cones of depression near Marion, Andrews, Georgetown, Pawleys Island, and Florence, S.C. (see Figure 1-2). The US Geological Survey (USGS) reported the potentiometric decline that occurred between predevelopment (1926) and 1982 near Florence, and between predevelopment (1879) and 1989 near Charleston, focused near the Mount Pleasant area, which used the Middendorf aquifer as a sole source of potable water as late as 1993. Groundwater flow for the period 1879–1989 was simulated for the Black Creek and the Middendorf aquifers and saw major declines from predevelopment times, with major cones of influence in major cities (see Figures 1-3 and 1-4).

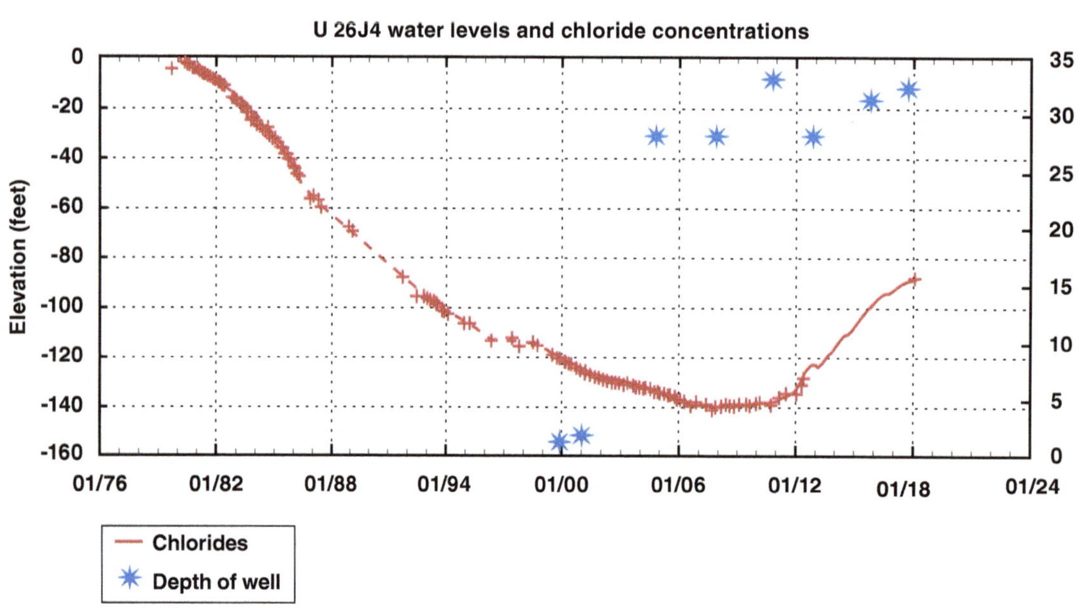

Source: NCDEQ 2018

Figure 1-1 Decline in the Black Creek aquifer in rural Jones County, N.C.

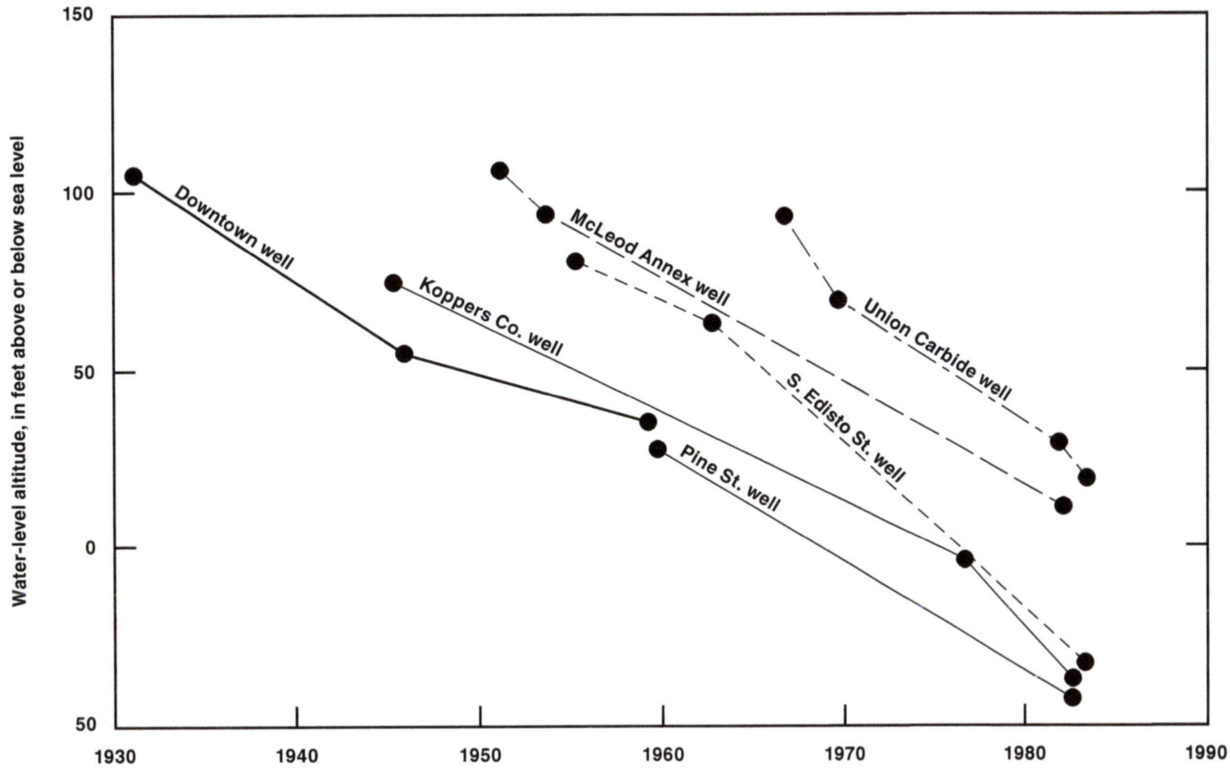

Source: Modified from Aucott and Speiran, 1985a

Figure 1-2 Declines in the aquifer formations near Florence, S.C.

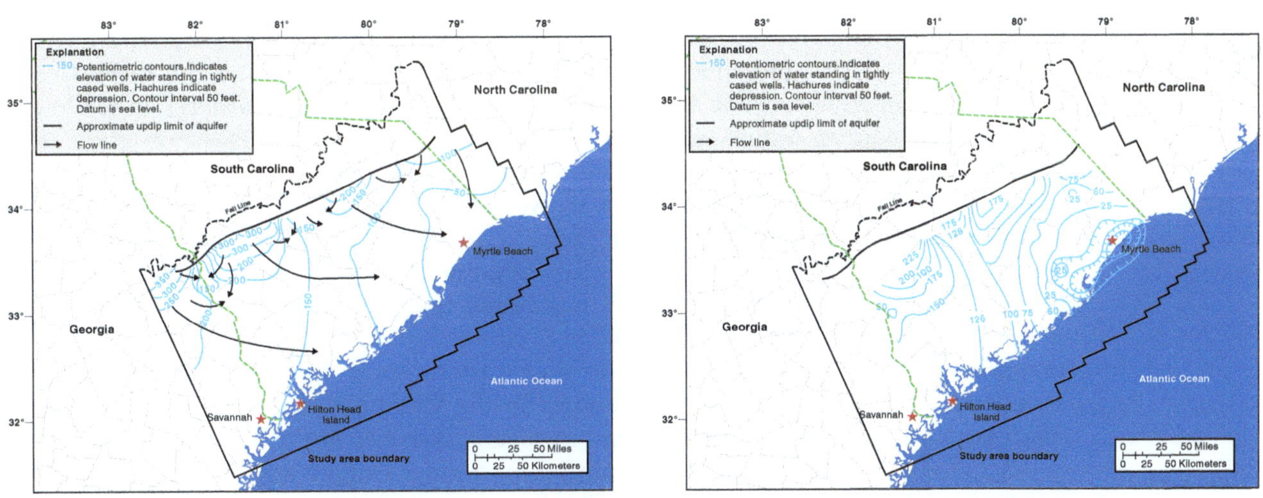

Source: Aucott 1996

Figure 1-3 Pre- and post-development (1982) head, Black Creek aquifer

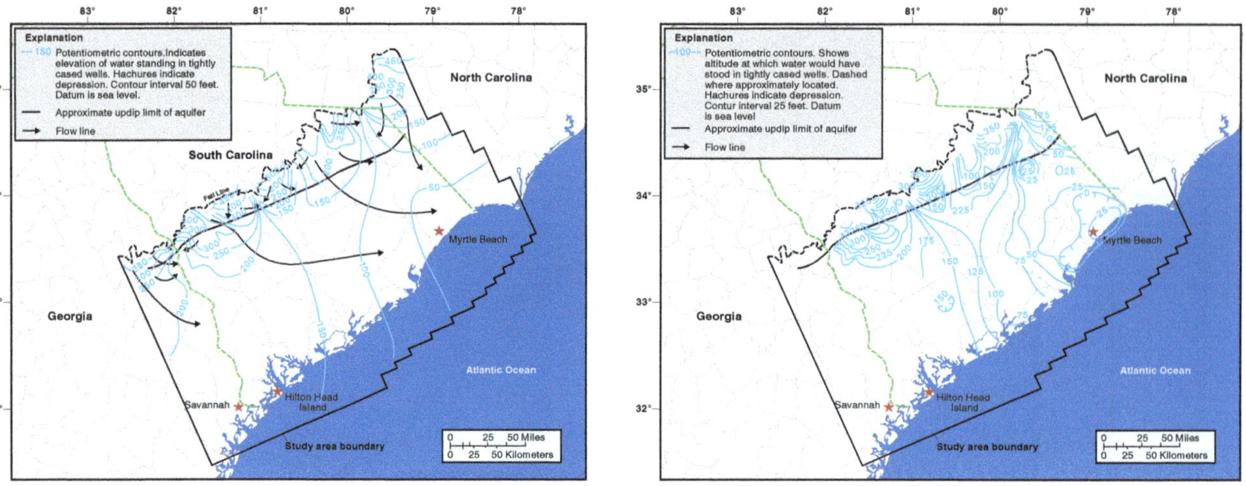

Source: Aucott 1996

Figure 1-4 Estimated pre- and post-development (1982) of the Middendorf Aquifer

The problem was that the declines were noted but basically ignored by the water users in the area. No user had a groundwater model in the 1980s, and no user appears to have acted to address the declining aquifer issue through changes in withdrawal patterns; they just lowered the pumps. No groundwater management plans were developed by the utilities, although the North Carolina utilities knew a shallower formation (the Castle Hayne) was available, but they had historically avoided it because its use would require lime softening treatment (as Camp LeJeune has done for more than 50 years). The decline in the aquifer levels was attributed to drought, competition, and other things, but not water use by the utilities. Regulatory action was required.

The declines became more acute in the 1990s. Because of these declines, the resulting increased potential for lateral saltwater migration along the coast and upward leakage of brackish water from deeper aquifers prompted the North Carolina Division of Water Resources (DWR) to institute the Central Coastal Plain Capacity Use Area (CCPCUA) rules for 15 counties, effective August 1, 2002 (North Carolina Department of Environmental Resources 2001). The guidelines required water users to cut back on the Black Creek aquifer and find other sources. As a result, water suppliers in North Carolina shifted to the Castle Hayne aquifer and decreased use of the Black Creek aquifer. However, the Castle Hayne is a shallower formation and is susceptible to saltwater intrusion.

South Carolina had similar issues. However, the Castle Hayne aquifer formation is lacking in South Carolina, so most of the South Carolina communities switched to surface waters as a result of aquifer declines. Fortunately, they were available to most of the larger utilities. Others, such as Myrtle Beach, monitored the decrease in its water levels until it combined its system with a consolidated Grand Strand Water and Sewer Authority that serves Horry County and others in northeastern South Carolina. Because water is a regional issue, many of the communities in the Carolinas have created countywide districts to address water supply issues. In several counties in the area, the Black Creek and Middendorf are being used or explored for aquifer storage and recovery wells. The good news: alternatives existed for the Carolinas.

GOAL OF THE MANUAL

The coastal Carolinas are not the only place in the United States that has ongoing issues with aquifer drawdowns and lacked groundwater management plans. Unlike the expectations of many who thought that the Southwest would have the first widespread aquifer to be drained, this water sustainability crisis was in the eastern United States. Fortunately, alternative solutions existed. But what about the central and western portions of the country? Finding out where at-risk water users exist and creating a pathway to them to determine the best means to manage water supplies sustainably is the goal of this manual. As will be outlined, the issues are more acute in the central and western states. The options to move to surface waters may not be available like it was in South Carolina, creating major consequences for local economies. The issue of managing groundwater supplies is critical in light of the identification of areas of the United States that have experienced aquifer overdrafting/mining, competition for the same source, and regulatory hurdles that do not protect groundwater sources for the future. A key component is the concept that planning the utilization of water supplies requires understanding how the hydrologic cycle provides water to the groundwater service area (e.g., recharge basin), in what quantities, and with what reliability. Reliability is a risk issue; is the precipitation consistent or are there significant fluctuations that disrupt ongoing basin development? From a practical perspective, sustainable development generally means addressing the environmental, economic, and social concerns underlying the sufficient availability of water for the locale.

The AWWA's Groundwater Resources Committee emphasizes the need for utilities to understand the effects of groundwater pumpage and the long-term impacts it has on the local community, the future economic conditions, and sustainability for local consumers. This need is a major extension of AWWA's core mission to provide guidance to policymakers, the water sector and professionals on how to deal with the restrictions in water supplies, and to develop more accurate measurements of available supplies; create better tools to evaluate (and reduce) water mining; and create management, policy, and technology solutions to address the problem of water shortages. This concept is fully in line with the mission of the Water Resource Sustainability Division; meets one of the top issues facing the groundwater sector; aligns with active professionals in the field; and can attract planners, geologists, administrators, and engineers to the discussion.

M73 is intended to outline a platform of key requirements, discuss existing groundwater management tools, and provide an overview for utilities of the data needed to create computer models of aquifers that are or in the future may be needed as a water source. The goal is to help utilities that may be subject to overdrafting or have groundwater quality concerns identify means to avoid further damage to aquifer systems, i.e., limit the future potential for a Carolinas-type problem. Toward these goals, the following chapters address:

- Chapter 2 – The Occurrence and Behavior of Groundwater;
- Chapter 3 – Motivation for Groundwater Management;
- Chapter 4 – Groundwater/Aquifer Characterization Tools;
- Chapter 5 – Monitoring to Support Management of Regional Groundwater Systems;
- Chapter 6 – Groundwater Models;
- Chapter 7 – Case Studies Using Groundwater Models (modeling case studies around the United States that demonstrate features and issues relevant to water utilities with groundwater as source water); and
- Chapter 8 – The Future of Groundwater Management.

REFERENCES

Aucott, W.R. 1996. *Hydrology of the Southeastern Coastal Plain Aquifer System in South Carolina and Parts of Georgia and North Carolina*. Reston, Va.: USGS.

Aucott, W.R. and Speiran, G.K. 1985a. Ground-water Flow in the Coastal Plain Aquifers of South Carolina. *Groundwater*. 23:6:736–745.

Aucott, W.R. and Speiran, G.K. 1985b. *Potentiometric Surfaces of November 1982 and Declines in the Potentiometric Surfaces Between the Period Prior to Development and November 1982 for the Coastal Plain Aquifers of South Carolina*. Reston, Va.: United States Geological Survey (USGS).

Cooke, C.W. 1936. *Geology of the Coastal Plain of South Carolina*. Washington: USGS.

Hockensmith, B.L. 1997. *Potentiometric Surface of the Black Creek Aquifer in South Carolina, November 1996*. Columbia, S.C.: South Carolina Department of Natural Resources.

Hockensmith, B.L. 2003a. *Potentiometric Surface of the Black Creek Aquifer in South Carolina, November 2001*. Columbia, S.C.: South Carolina Department of Natural Resources.

Hockensmith, B.L. 2003b. *Potentiometric Surface of the Middendorf Aquifer in South Carolina, November 2001*. Columbia, S.C.: South Carolina Department of Natural Resources.

Hockensmith, B.L. 2008. *Potentiometric Surface of the Black Creek Aquifer in South Carolina, November 2004*. Columbia, S.C.: South Carolina Department of Natural Resources.

Lautier, J.C. 2001. *Hydrogeologic Framework and Ground Water Conditions in the North Carolina Southern Coastal Plain*. Raleigh, N.C.: North Carolina Department of Environment and Natural Resources, Division of Water Resources.

NCDENR (North Carolina Department of Environment and Natural Resources). 2008. *North Carolina Aquifers*. Raleigh, N.C.: NCDENR, Division of Water Resources.

NCDEQ (North Carolina Department of Environmental Quality). 2018. *Central Coastal Plain Capacity Use Area Assessment Report*. Raleigh, N.C.: NCDEQ, Division of Water Resources.

North Carolina Department of Environmental Resources. 2001. *Central Coastal Plain Capacity Use Area Rules*. Raleigh, N.C.: NCDENR, Division of Water Resources.

Stringfield, W.J. and Campbell, B.G. 1993. *Potentiometric Surfaces of November 1989 and Declines in the Potentiometric Surfaces between November 1982 and November 1989 for the Black Creek and Middendorf Aquifers in South Carolina*. Reston, Va.: USGS.

Winner, M.D. Jr. and Coble, R.W. 1996. *Hydrogeologic Framework of the North Carolina Coastal Plain, Regional Aquifer-System Analysis—Northern Atlantic Coastal Plain*. Reston, Va.: USGS.

Chapter 2

The Occurrence and Behavior of Groundwater

More than a third of the people served with public water supplies in the United States and Canada obtain their water supplies from groundwater (AWWA 2014). This demand is increasing as the number of small systems has grown from less than 5,000 to more than 40,000 since 1960 (Bloetscher 2019). In part, the expansion of groundwater-supplied utilities results from the lack of suitable sites to build impoundments to capture surface waters and the water quality of existing surface waters. Also, wells can be installed in small increments with little apparent impact per well. As areas grow, those small impacts become many wells that can create significant impacts on groundwater levels, but this may be overlooked by most users except when they need to lower their pumps. Because groundwater is not visible from the surface, the understanding of its behavior and occurrence among the public is limited. This chapter is intended to provide a general overview of:

- the hydrologic cycle and general groundwater concepts;
- major conditions that impact groundwater;
- climate impacts on groundwater;
- sustainability, development, and urbanization; and
- water quality.

HYDROLOGIC CYCLE

Saltwater in oceans makes up 97.3 percent of all water in the world. For most people, this water is not useful for water supply purposes because of the cost of treatment or the distance to convey desalinated seawater. Of the remaining 2.7 percent freshwater, more than two-thirds exists as ice in the polar caps while the rest (0.61 percent of all water) is freshwater in lakes, rivers, streams, and groundwater. More than half may be groundwater (AWWA 2014).

Water is not stagnant, whether in the ground or above. Water movement above, on, and below the earth's surface is defined as the hydrologic cycle (Figure 2-1). The hydrologic cycle is the main concept used in the development and management of water supplies. The components of the hydrologic cycle are:

- evapotranspiration (ET), which is the combination of evaporation from open water and transpiration from plants as an average over an area,
- precipitation,
- surface water and runoff,
- infiltration and percolation to the subsurface, and
- groundwater recharge and discharge.

Evaporation and Transpiration

Although the hydrologic cycle is continuous, evaporation from open water and transpiration from plants are major drivers created by atmospheric conditions. Evaporation and transpiration, commonly combined and referred to as the process of ET, is water vapor entering the atmosphere. ET rates vary, depending largely on the amount of solar radiation, the elevation, the latitude of the catchment area, the amount of heat, water surface area, wind, humidity, time of year, and vegetative cover. Areas close to the equator tend to have higher ET rates. Figure 2-2 is a map of ET rates in the United States. Figure 2-2 shows that ET rates vary widely across North America and throughout the year, and whether there is open water to evaporate (hence, the desert Southwest does not have a lot of ET because there is very little surface water). In addition, visitors to the Rocky Mountains are often surprised how ET causes new snow to disappear despite the temperature being below freezing (a process called "sublimation").

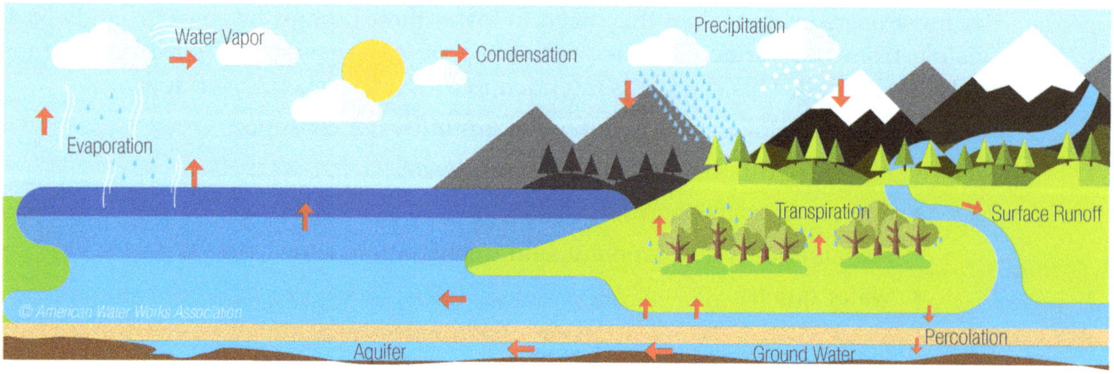

Source: DrinkTap.org by AWWA

Figure 2-1 Hydrologic cycle

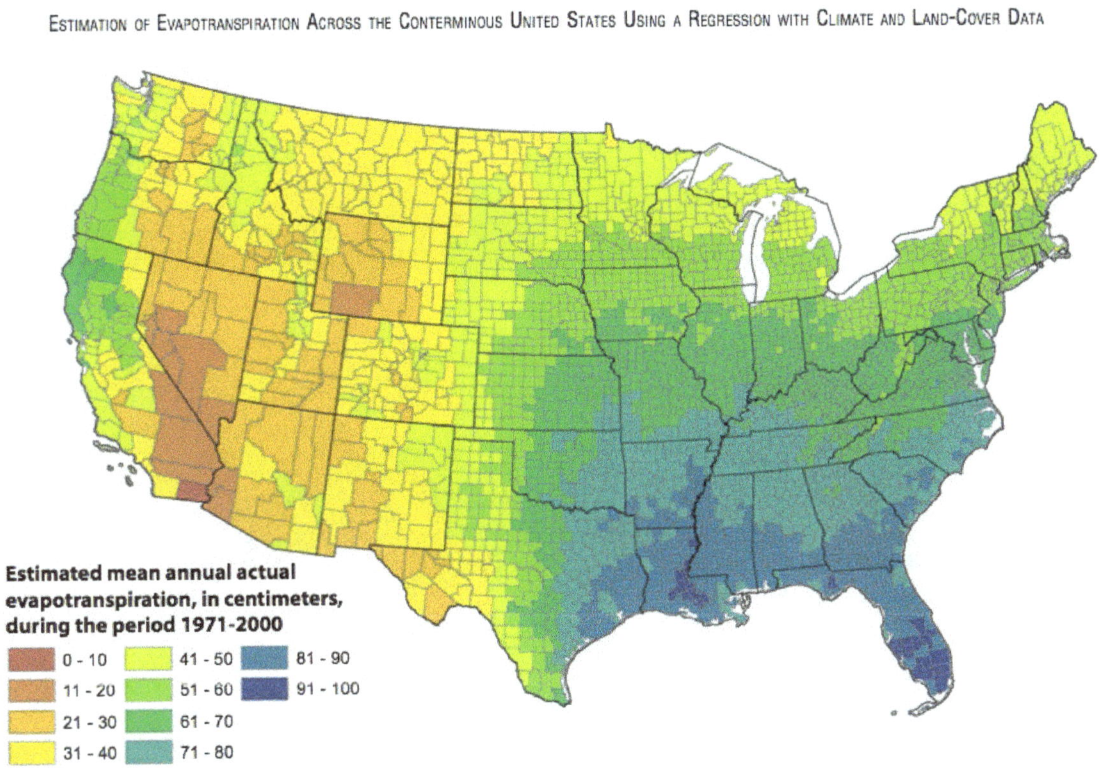

Source: National Oceanic and Atmospheric Administration 2013

Figure 2-2 Evapotranspiration rates

In subtropical areas during the wet season or during summer months in northern latitudes, large bodies of water, including wetlands, lakes, and estuarine areas, have high evaporation rates. This rising moisture forms clouds that condense and return the water to the land surface or oceans in the form of precipitation. The highest evaporation rates are associated with shallow, open water bodies. Water as much as 4 ft below the surface may be subject to evaporation to some degree.

Precipitation

Figure 2-3 shows average annual precipitation in the United States. As water vapor enters the atmosphere, conditions create the need for the atmosphere to release some of this moisture as precipitation. Precipitation occurs in several forms, including rain, snow, sleet, and hail. The intensity of the resulting rainfall is a major area of hydrologic study as it seems rainfall intensity is increasing with time. Rainfall intensity can vary up to 5 in./h in subtropical areas but is commonly 2–3 in./h across North America. Rainfall intensity is relevant because the rain wets vegetation and other surfaces, then infiltrates into the ground or travels toward shallow groundwater bodies or toward areas of lower ground elevation. Infiltration rates vary widely, depending on land use, development, the character and moisture content of the soil, and the intensity and duration of the precipitation event. Infiltration rates can vary from as much as 1 in./h in mature forests on sandy soils to almost nothing in clay soils and paved areas. If the rate of precipitation exceeds the rate of infiltration, overland flow (or runoff) occurs. This is why development and deforestation are critical concerns in watersheds as both diminish infiltration potential and increase the potential for (often faster) runoff. As development alters the land surface characteristics, stormwater systems are often overloaded and groundwater does not recharge because

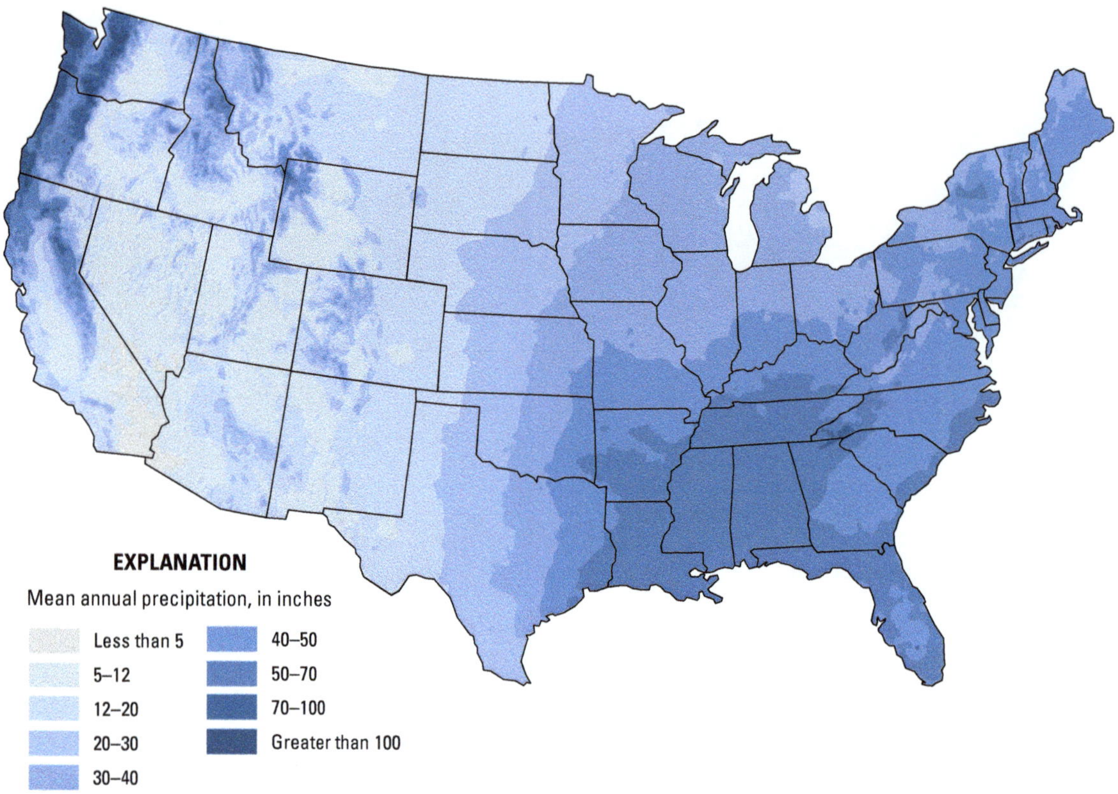

Source: Reilly et al. 2008

Figure 2-3 Mean annual precipitation for the conterminous United States, 1890–2002

water stays on the ground for shorter periods of time—a major concern for water purveyors who rely on groundwater as their supply source.

Surface Water and Runoff

Precipitation that runs off the land, reaching streams, rivers, or lakes, or groundwater that discharges into these water bodies is surface water. While recharge from the surface precipitation is the norm for groundwater, there is the potential for surface water bodies to recharge groundwater. This surface water recharge of groundwater can be more important in more arid areas such as the West where the groundwater is deeper than the surface waters. The reverse is true in more humid areas; the propensity is much higher for groundwater to flow toward water bodies than for water bodies to recharge groundwater. As a result, rivers and canals can act as "sinks" to lower connected groundwater levels in lower, more humid areas. South Florida is an example of this. Surface water bodies that mix with saltwater bodies along the coast are called estuaries (e.g., where a river meets the ocean in a delta). Surface water flow is controlled by the topography because water on the surface flows downward by gravity, which means it generally and eventually flows toward the oceans.

Infiltration and Recharge

Precipitation that percolates downward through porous surface soils is the primary source of water for groundwater. Those areas on the surface that create downward flow are called "recharge areas." The characteristics of soil material depend on the climate, soil chemistry, the types of organisms in and on top of the soil, the topography of the land, and the amount of time these factors have acted on the material.

Soil types also play a role in determining plant distribution (AWWA 2014). Soil has the capacity to absorb some moisture initially, a factor called "initial infiltration." Initial infiltration fills air pockets in the root or plant zones or, in some lexicons, the *unsaturated zone*. Because of the variability of permeability and transmissivity of different soils, the rate of groundwater recharge from precipitation will vary.

Below the unsaturated soil zone, all rocks (including unconsolidated sediments) under the earth's surface can be classified either as aquifers, semi-confining units, or confining units. An aquifer is a rock formation that will yield water in a usable quantity to a well or spring. Because groundwater moves slowly, some groundwater has been stored in aquifers for hundreds or thousands of years. The longer the water remains in contact with the rock, the more constituents the water may contain because of added contact time to dissolve the formation, although flow velocity also increases dissolution.

The ability of water to flow through the porous media is referred to as *hydraulic conductivity (K)*. A confining unit is a formation having very low hydraulic conductivity that restricts the movement of groundwater either into or out of adjacent rock formations as shown in Figure 2-4. Clay layers (and some rock formations such as shale), although having high porosity, have *low hydraulic conductivity* and can be functionally impermeable. As a result, engineers and hydrogeologists typically look for limestone, sandstone, and alluvial formations, and in some areas, fractured bedrock, as these formations tend to have high hydraulic conductivity. They avoid clay, shale, and similar formations, which have low transmissivity for water supply development.

Groundwater occurs in aquifers that are defined as either "confined" or "unconfined." If the aquifer is near the land surface and rainfall will impact the water level (filling it), such aquifers are termed *unconfined* aquifers. The upper surface of the saturated zone is free to rise and decline in direct relation to recharge by precipitation. Wells that pump water out of unconfined aquifers are water-table wells. The water level in these wells generally indicates the position of the water table in the surrounding aquifer. With unconfined aquifers, local rainfall and snow melt will recharge the formation regularly. This is not the case with confined aquifers.

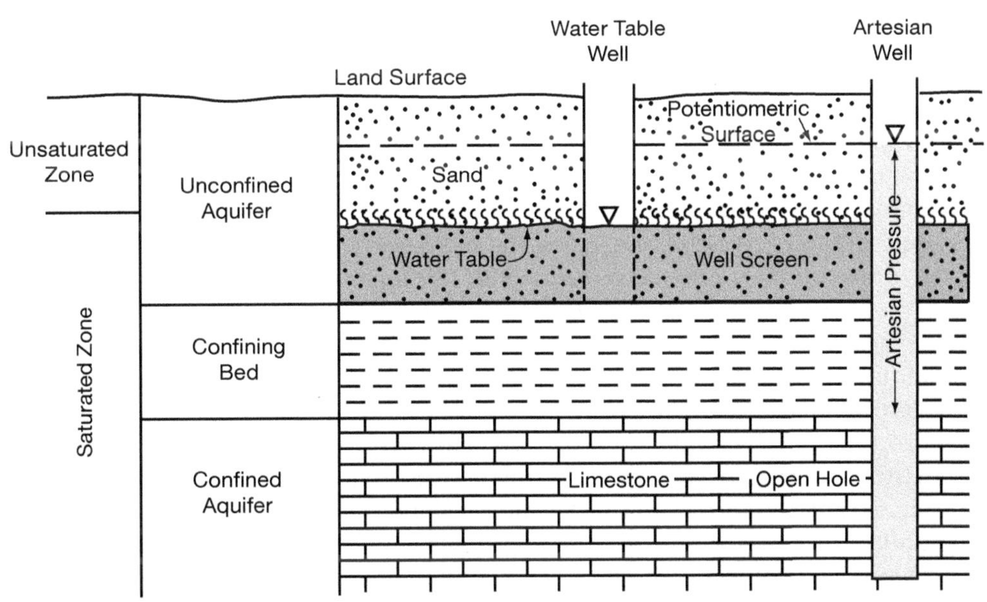

Source: M21 (AWWA 2014)

Figure 2-4 Geologic configuration of aquifers and confining beds

Confining layers or confining units are formations that water cannot flow through easily. Clays are a good example. Below the confining unit is a confined aquifer (or aquifers). A *confined* aquifer is not locally open to the surface because of the confining formation. In confined aquifers, water completely fills the formation. Such aquifers may be referred to as artesian aquifers depending on whether the water level rises above the top of the aquifer formation. If the water level in an artesian well stands above the land surface, the well is a flowing artesian well. How far the water moves above the top of the formation is head or the pressure the formation is under.

Recharge areas for confined aquifers can be located far from the point of use. Hence, snow melt and rainfall occurring some distance away—maybe hundreds of miles in some cases—is what recharges the local confined aquifer. The head within the formation may provide some guidance on recharge areas as it will indicate the approximate elevation of the recharge area. With respect to water quality, for a confined aquifer to contain fresh water, there must be a source of recharge, some degree of flow (albeit slow), and a discharge area (to cause the flow). If there is no recharge or movement, the aquifer will often become brackish through dissolution of the minerals in the rock because of contact time.

Note that the amount of water available in the surficial aquifer will vary as the water table changes. The impacts are less obvious in confined aquifers where pressure changes will need to be monitored to determine how much water is available. Too many utilities simply lower the pumps in confined aquifers, creating a long-term decrease in available water and potential for denuding (or dewatering) the aquifer in time. This aquifer degradation happened in the eastern Carolinas in the mid-1980s.

When a well is pumped, the level of the aquifer falls as the water immediately surrounding the well is drawn through the pumping well. The lower aquifer level is called "drawdown," and the 3-D cross section is called the "cone of depression." If the aquifer transmits water easily (i.e., has high transmissivity), the drawdown may be slight and the cone of depression flat and widespread as depicted in Figure 2-5. For the same withdrawal, if the aquifer has low hydraulic conductivity, drawdown may be significant, and the cone of depression steep. Hence, the type of rock strata that underlies the area may have significant bearing on the quantity of groundwater that can be produced (AWWA 2014). Withdrawal flows need to be adjusted to address concerns with drawdown and the extent of the cone of depression.

Under natural conditions, groundwater moves downgradient until it reaches the land surface at a spring or through a seep along the side or bottom of a stream channel or estuary, or, in deeper aquifers, the ocean. Groundwater in the shallowest part of the saturated zone moves from interstream areas toward streams or the coast. In many areas, the direction of groundwater movement can be derived from observations of land topography when the land slopes toward water bodies. Thus, the water table aquifer levels usually replicate the land surface as shown in Figure 2-6. Changes to upstream land use can alter confined aquifer yields, typically lessening the recharge and causing the pressure in the formation to diminish. This pressure is termed the "potentiometric" head.

A common assumption made by engineers and hydrogeologists is that water moves through confining units to some degree. Many formations may be semi-confining units indicating water movement to some degree, but a true confining unit permits no passage. Vertical migration as vertical movement is a function of both the geometry and hydraulic properties of the semi-confining layer. However, a mistake in many modeling efforts is to assume migration. One should never assume that the connection between two aquifers is significant because groundwater moves slowly, and even more slowly through confining units. The existing tests used to measure the potential for movement are limited as well. Hence, vertical recharge to confined aquifers through a confining unit may be minimal in many formations. A modeling exercise by Gao (2010) showed that vertical recharge may be estimated to be more than 1,000 times lower than the estimates used by hydrogeologists.

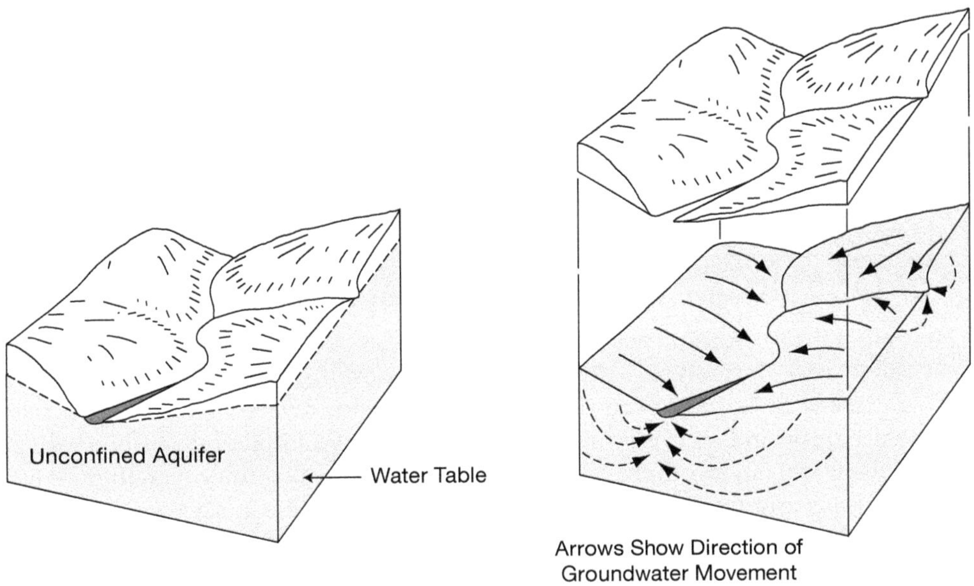

Source: M21 (AWWA 2014)

Figure 2-5 Development of a cone of depression, showing the central well and a series of monitoring wells that were used to define the cone of depression.

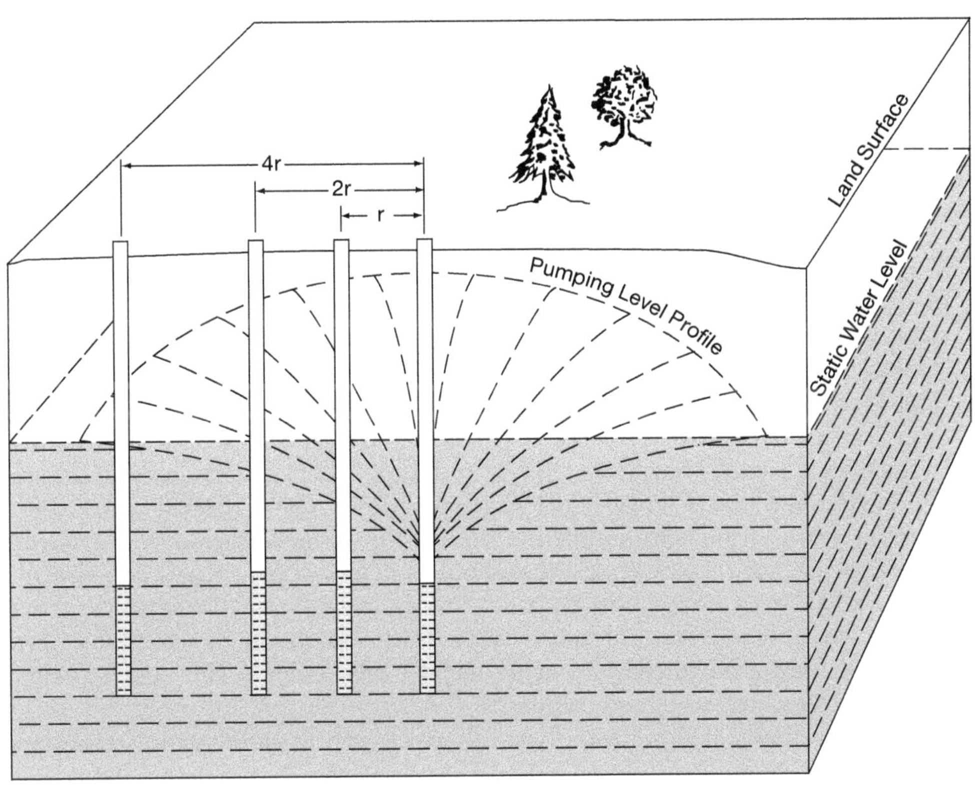

Source: M21 (AWWA 2014)

Figure 2-6 Groundwater levels typically mimic the surface topography in a given watershed

Therefore, utilities and engineers modeling wellfields should take note that vertical recharge may be zero through a thick, continuous confining layer.

In the western part of the contiguous United States, conditions are more variable than those described above. In this area, streams flowing from mountain ranges onto alluvial plains lose water to the alluvial deposits. Thus, groundwater in the upper part of the saturated zone flows down the valleys and away from the streams.

THE IMPACT OF LAND SUBSIDENCE ON GROUNDWATER

A significant consequence of groundwater development from unconsolidated or highly friable rock formations can be the downward movement of the land surface called *subsidence*. Subsidence can occur when the groundwater, which exerts pressure on the adjacent soil and rock, is removed, relieving the interstitial pressure between grains created by the water (an incompressible liquid). With removal of the water, there is nothing to hold the grains apart, so the formation collapses, causing the surface topography to be altered. The collapse can cause the reorientation of aquifer grains, particularly unconsolidated sediments. Recharging an aquifer that has collapsed will not restore the land surface to its prior state because the collapsed formation has less void space (which is why the collapse happened in the first place). The recharge capacity may be lowered significantly, and the ability to accept recharge compromised. Development of groundwater should consider possible land-surface subsidence, especially if subsidence has been previously documented. Areas with considerable subsidence may exacerbate flooding in flood-prone areas since the land is lower. Unfortunately, the information needed is not available in sufficient detail in most areas.

CLIMATE IMPACTS ON GROUNDWATER

There has been significant discussion in popular media on the topic of climate change with much focus on the potential for greenhouse gases to be the causal agent for the accelerating warming trend on Earth. While there are several causes that may contribute to changes in the worldwide climate—including greenhouse gases (IPCC 2007), sun spot activity (Singer 2008), land use changes (Marshall et al. 2004), and even a coming change in the pole location (Hapgood 1999)—the reality is that, regardless of the cause, much of the scientific community agrees that climate change *is* occurring and that it may portend a change in water availability in the near future (Bloetscher et al. 2012, Heimlich et al. 2009, Bloetscher 2008, and references therein).

What is happening appears to be occurring at a faster rate than suggested by studying the past. The Earth's climate undergoes constant changes, but these changes have occurred at a much slower, geologic timescale. A series of continental glaciers extended as far south as the 40th parallel in North America and Eurasia in the past. The ebb and flow of glaciers has occurred throughout the *Pleistocene* geologic era, which started more than 600,000 years ago and extending to as recently as 10,000 years ago. These glacial advances and retreats profoundly altered the landscape, e.g., covering northern Europe and the northern United States with deep till and outwash soils and leaving behind large, productive, sand-and-gravel aquifers.

Less than 5,000 years ago, the Sahara Desert was a thriving, water-soaked area that supported significant human population; however, the climate changed, and the area has become a huge expanse of sand since then. Much of the groundwater available in the western United States is fossil water from the past 600,000 years. Brief interludes of warmer and cooler periods occurred during 400–1300 AD. However, in each case, the factors creating the changes in climate are unclear (AWWA 2014), and man was unable to adapt to many of these changes. The main concern for utilities is that climatic variations alter the hydrologic cycle, and that the current data indicates that the hydrological cycle is already being impacted: more intense rainfall, less recharge, less snowpack, and higher seas (Dragoni and Sukhija 2008; IPCC 2007; Huntington 2006; Labat et al. 2004; Buffoni 2002; and Dragoni 1998), all of which may mean less freshwater recharge to existing aquifers. The uncertainty associated with temperature and precipitation yields higher uncertainty when translating rainfall and temperature to changes in ET, runoff, and aquifer recharge (Dragoni and Sukhija 2008; Di Matteo and Dragoni 2006; Strzepek and Yates 1997). That means the sustainability of supplies may be overestimated using assumptions based on historical data.

SUSTAINABILITY

With sustainable water sources, the key component is that planning the utilization of water supplies is required to determine how the hydrologic cycle provides water to the service area (e.g., recharge basin), in what quantities, and with what reliability. The reliability is a risk issue; is the precipitation consistent or are there significant fluctuations that disrupt ongoing basin development (Molak 1997)? Additionally, because of the variability of groundwater uses and demands over time, each basin has unique characteristics (Bloetscher and Muniz 2008). From a hydrologic perspective, the term "sustainable yield" is the amount of water that can be withdrawn from a source at rates that are less than their recharge potential and that do not diminish the quantity or quality of the source or basin. While many areas attempt to develop sustainably, many others have not. The eastern Carolinas example in Chapter 1 is but one example of unsustainable groundwater development. Economics, legal ramifications and property rights, a lack of regulations, and public perception are all factors that delay action.

While no one wants to create a water supply crisis, the groundwater supply is being overdrafted where there are increasing drawdowns, indicating that doing what has been done in the past is not a sustainable usage rate. There is suspicion among many groundwater professionals and regulators that most confined aquifers are *not* sustainable supplies by the very definition of the term "confined." Within regionally sustainable situations, unsustainable pumping can occur locally. For example, much of the western groundwater appears to be unsustainable in the long term because of limited rainfall and very low recharge rates locally (Reilly et al. 2008). One problem is competition, a situation highlighted in the AWWA compendium from 2010. Because each sector of water users is focused on their needs, no sector knows much about the others; where there are potential restrictions in water use, they see each other as competitors in a water-limited market. Hence, solutions are often litigated as opposed to resolved amicably between parties to the benefit of the public good.

Typically, there are a variety of uses competing for water resources, and each basin has unique characteristics for the following (Bloetscher and Muniz 2008):

- agriculture,
- ecosystems,
- urban demands,
- industrial demands, and
- cooling water for power generation.

From a practical perspective, sustainable development generally means addressing the environmental, economic, and social concerns over the long term, which may indicate that short-term needs are not part of long-term solutions. Researchers can define a comprehensible concept of "sustainability," but practitioners emphasize feasibility and limitations to

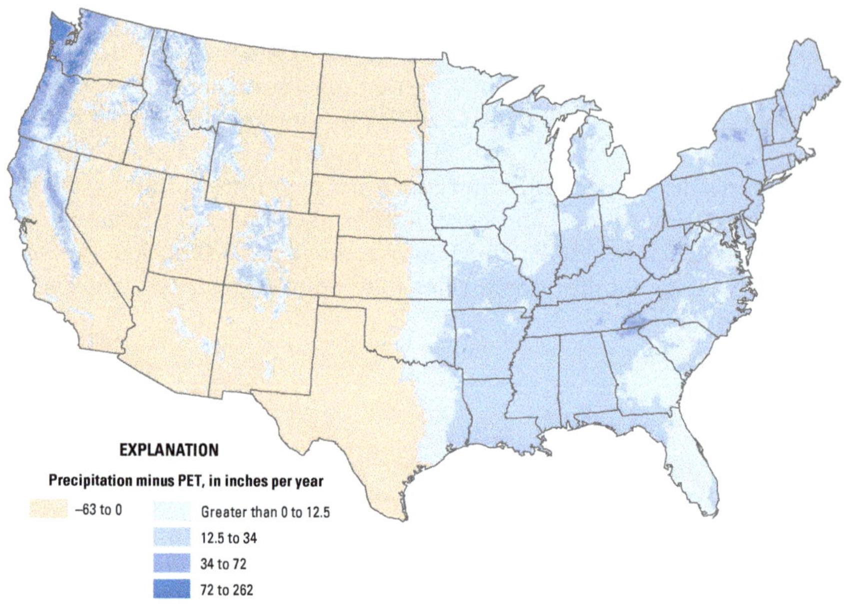

Source: Reilly et al. 2008

Figure 2-7 Difference between average annual precipitation and potential ET (PET) rates

sustainability of the ecological, economic, social, and industrial community (Starkl and Brunner 2004). Scanlon et al. (2005) identified the need to fully optimize management of water resources, including those for ecosystem purposes in as many basins as possible because the ecosystem drives the economic activity (tourism, fishing, etc.). Most surficial changes decrease available recharge to groundwater.

Although determining the amount of groundwater available may seem straightforward, it is quite complex as noted by Reilly et al. (2008). Reilly et al. (2008) outlined the condition of groundwater in the United States and found that the loss of groundwater supplies is because of overpumping in many areas. In part, this was because, while eastern rainfall is much higher than the west as noted in Figure 2-3, much of the Great Plains, Rocky Mountain, and Western states have ET rates that exceed precipitation. The result is Figure 2-7, which shows the difference between rainfall and potential ET rates. The red areas are areas where the ET rate is higher than the rainfall, meaning net rainfall (Rainfall – ET) for crops and other purposes is not available and extensive surficial water use is not a sustainable practice. This is confirmed by the fact that in many of these areas, streamflows are variable and limited, so groundwater is used to ensure water supplies for crops and people. Aerial views that show extensive irrigation "crop circles" in many areas of high agricultural production and low rainfall; this is generally an example of unsustainable water use. Figure 2-8 shows that many of these areas are water deficit areas, meaning that there is a deficit of available of water in the area as determined by US Geological Survey (USGS) data. Figure 2-9 shows the amount of water available for recharge throughout the United States. Most areas have less than 1 in. per year available for recharge, yet aquifers are falling in feet per year. Figure 2-10 combines regional water-level declines and local water-level declines for changes on a national scale. The red regions indicate areas in excess of 500 mi^2 that have water-level decline in excess of 40 ft in at least one *confined* aquifer since predevelopment, or in excess of 25 ft of decline in unconfined aquifers since predevelopment. The confined aquifers recharge at very low rates, meaning that in most cases, these aquifers are being mined (overdrafted). The blue dots on Figure 2-10 are wells in the USGS National Water Information System database where the measured water-level decline over time is equal to or greater than 40 ft.

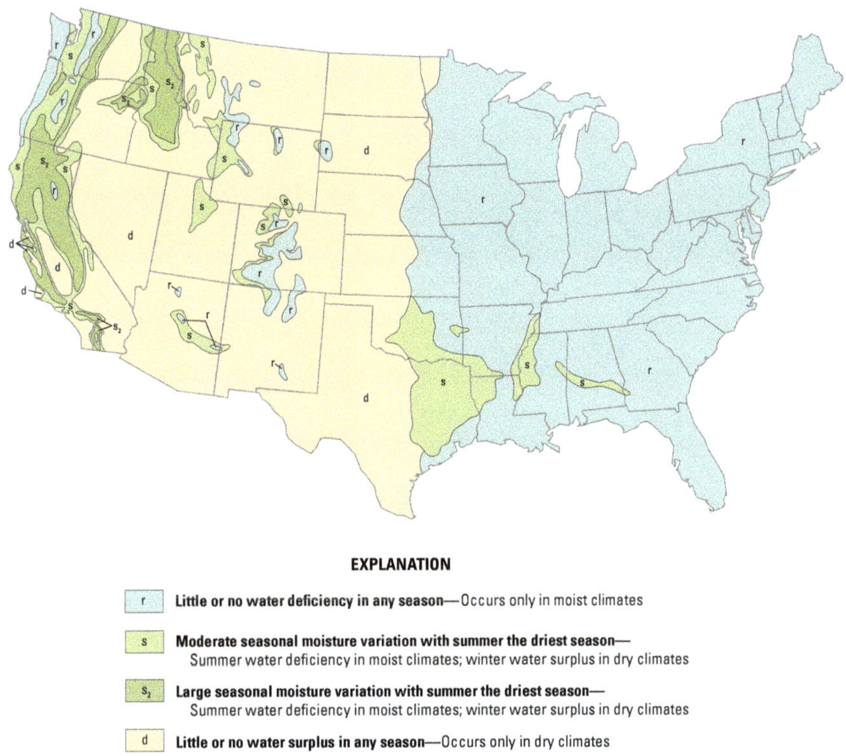

Source: Thornthwaite 1948

Figure 2-8 Water deficit areas

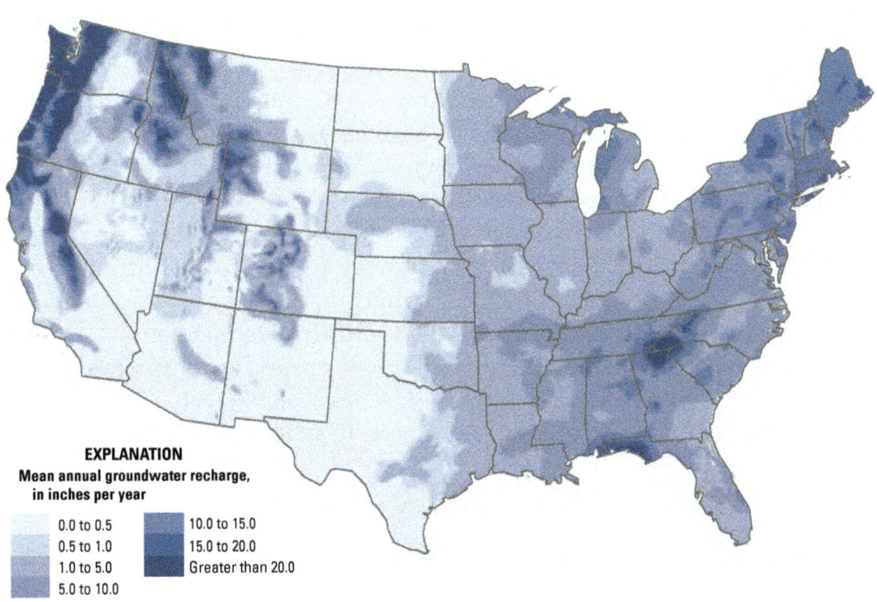

Source: Reilly et al. 2008; Wolock 2007

Figure 2-9 Water available for recharge throughout the United States (Note: Most areas are very low.)

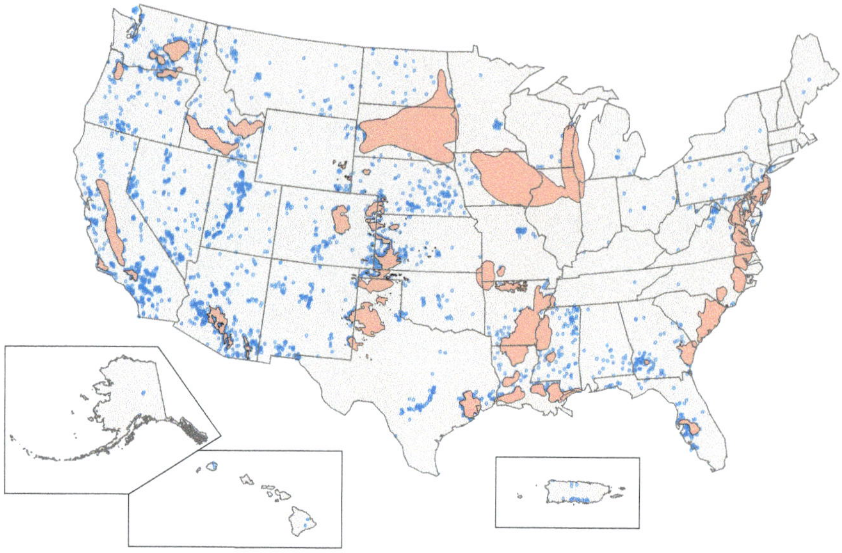

Figure 12. Water-level declines. Red regions indicate areas in excess of 500 square miles that have water-level decline in excess of 40 feet in at least one confined aquifer since predevelopment, or in excess of 25 feet of decline in unconfined aquifers since predevelopment. Blue dots are wells in the USGS National Water Information System database where the measured water-level difference over time is equal to or greater than 40 feet.

Source: Reilly et al. 2008

Figure 2-10 Water-level declines

USGS reports the need for a nationwide effort to organize available information on changes in groundwater storage, similar to what was done for the High Plains aquifer (Reilly et al. 2008). Confounding the problem is that many aquifer systems cross political boundaries, so careful regulation in one jurisdiction may not be supported by others.

Regulations to ensure groundwater supplies are sustainable are recent. Since 1974, Florida has had a groundwater permitting program in place for large agricultural and utility entities. Over time, this program has become more sophisticated given increasing demands in the state so that today, utilities must demonstrate via modeling that their withdrawals will not affect other users. California passed the Sustainable Groundwater Management Act in September 2014 with a goal to ensure better local and regional management of groundwater use; it seeks to have a sustainable groundwater management in California by 2042. The basic tenets were to ensure that any use of the groundwater be both reasonable and beneficial, while authorizing local agencies to implement a groundwater plan, and authorizing the California Department of Water Resources (DWR) to offer support to "entities that extract or use groundwater to promote water conservation and protect groundwater resources" (DANR 2018, CDWR 2017, California Senate 2014). Texas has begun efforts to legislate means to manage groundwater as much of the state depends on it.

DEVELOPMENT AND URBANIZATION IMPACTS

Groundwater recharge is affected by precipitation, actual ET, topography, land use, soil type, land cover, aquifer transmissivity, vegetation characteristics, and contributions to recharge along active stream channels (Herrera-Pantoja and Hiscock 2008). In most cases, there is limited historical information about the quantity of water that was initially available, so while one can recognize that historical water availability in a basin changed as a result of water use practices, the magnitude of the changes is uncertain.

The change in land use from forests to agriculture or urban uses can have significant impacts on runoff characteristics. Hydrologists point out that areas with higher surface imperviousness will accelerate runoff and decrease infiltration. Urban land use increases runoff because of imperviousness from buildings, parking lots, roads, and other improvements that replace forest or grassland cover (Bloetscher and Muniz 2008). Examples of activities that may affect raw water supplies include delivery times of the water through piping installed to reduce flooding as opposed to natural channels and/or development that replaces irrigated agriculture with paving. Urbanization reduces recharge to groundwater but increases flows to surface reservoirs and streams with potential added nutrients and contaminants. As a result, determining the changes to groundwater availability involves more than calculating the volume of groundwater within an aquifer: it requires a consideration of recharge, water quality, the economics of recovery or the treatment for poor quality water, interconnectedness with the hydrologic system, and ecosystem/user demands. Groundwater management thus involves understanding adjacent users, future demands, limitations of the aquifer and formations and the impacts these have to water supply, and water quality. Water quality, the subject of Chapter 3, is another reason groundwater management is needed.

INTEGRATING GROUNDWATER MANAGEMENT

As this discussion has shown, groundwater has been and continues to be managed under distinctly different approaches by region, by state vs. federal rules, and by tradition. A more integrated management approach in the future is key to having groundwater as a water supply, as base flow to rivers, and as a fundamental basis for healthy systems. Integrated management is required as a part of utility management plans for the future as utilities grow and develop their raw water sources, as well as bring in alternative water sources. Those utilities that invest in their future now will be best placed to weather the uncertainties of future water supply and demands.

Of the two major concerns of groundwater management—quantity and quality—the former has traditionally been the reason that brought entities with an interest in water together. Three methods employed to stretch supplies further by expanding locally available supplies are artificial recharge, conjunctive use, and water wheeling. In arid regions such as the southwestern United States, artificial recharge has been practiced for decades, with conjunctive-use agreements initiated in the 1960s and water-wheeling arrangements starting in the 1990s.

Artificial Recharge

Artificial recharge augments groundwater supplies by supplementing natural recharge. Surface aquifers percolate stormwater captured from precipitation, imported water, and sometimes, recycled water. This concept is discussed more fully in the 4th edition of M21. It is applied in many parts of the country where there are large catchment areas that will percolate water easily. One example is Southeast Florida's water conservation areas.

Conjunctive Use

Groundwater management programs can improve the long-term reliability of water supplies by integrating the use of surface water and groundwater together or "conjunctively." Conjunctive use may be accomplished through either direct or replacement operations.

Direct conjunctive use places imported water into the aquifer, and the water is then pumped as needed. Replacement (or *in-lieu*) conjunctive use is when a water supplier uses imported water instead of pumping its groundwater, leaving the groundwater available in the aquifer for use at a later time. *Seasonal storage* operations recharge an aquifer with imported water during wet times of the year, when surface and imported waters are plentiful, and pump the water out during high-demand months, generally within that same water year. This management strategy results in a seasonal shift, with water typically stored in winter months and pumped during summer months. In some cases, it can be stored underground, the subject of AWWA Manual M63, Aquifer Storage and Recovery. *Carryover storage* occurs when the recharged water is held for use beyond the next water year, such as for a drought.

Wheeling

Water wheeling uses a water supplier's transportation system for delivery of water not owned or controlled by that agency. Wheeling is a common practice in the energy and communications industries. Wheeling permits water suppliers to move water supplies (groundwater or surface water) from areas of plenty to areas of need. Central Florida has a water wheeling program that withdraws water from wellfields that are unstressed to areas where stressed wells may have limited withdrawal allocations. Surface water, aquifer storage, and other concepts supplement the wheeling issue.

Some states have enacted legislation regulating wheeling. For example, the California Water Code requires that water wheeling must not harm any other legal user of water. Other states only require agreements among the parties and agencies that might permit supplies. The cost of wheeling typically recovers costs such as transportation, including capital and operating costs associated with pipelines, reservoirs, and other facilities required to distribute water.

Water Marketing and Water Transfers

Water marketing is the general term encompassing the transfer, lease, or sale of water or water rights. A water transfer is the shift or sale of water from a seller to a buyer. Water transfers can take many forms, each with its own benefits and risks. A *spot transfer* denotes purchase of supplies when needed, such as to offset the effects of a prolonged drought. Under an *option transfer*, the buyer takes a certain amount of water at any time during the life of the agreement, making option payments on an annual basis and making an additional payment in those years in which the water is needed. Buyers in a *core transfer* (or take-and-pay) take a specified amount of water each year and must pay the cost of that water, year in and year out, whether or not the water is needed. This concept is currently applicable only in the 18 western continental states where water rights can be bought and sold. In the remaining states, the waters are owned by the state. Water marketing is increasing in the general public's awareness because of concerns with transferring water from one basin to another, and potential issues that can develop because of decreased surface water and groundwater that is available to residents, natural habitats, and riparian corridors.

REFERENCES

AWWA. 2014 (4th ed.). M21, Groundwater. Denver, Colo.: AWWA.

AWWA. 2015. M63, Aquifer Storage and Recovery. Denver, Colo.: AWWA.

Bloetscher, F. 2019. *Infrastructure Management.* Plantation, Fla.: J. Ross Publishing.

Bloetscher, F.; Romah, T.; Berry, L.; Hernandez Hammer, N.; and Cahill, M.A. 2012. Identification of Physical Transportation Infrastructure Vulnerable to Sea Level Rise. *Journal of Sustainable Development.* 5:12:40.

Bloetscher, F. (ed.). 2010. *Sustainability Compendium.* Denver, Colo.: AWWA.

Bloetscher, F. 2008. The Potential Impact of Climate Change on Groundwater Recharge. Proc. GWPC Annual Forum Proceedings, Oklahoma City, Okla.

Bloetscher, F. and Muniz, A. 2008. Water Supply in South Florida—The New Limitations. Proc. AWWA 2008 Sustainable Water Sources Conf., Reno, Nev.

Buffoni, L.B. 2002. Variazioni Climate Change and Groundwater: A Short Review 7. *Italia Negli Ultimi 130 Anni. Bollettino Geofisico Anno XXIII.*

California Senate. 2014. *SB-1319, Groundwater.*

CDWR (California Division of Water Resources). 2017. *Sustainable Groundwater Management Act.* Sacramento, Calif.: CDWR.

DANR (University of California Division of Agriculture and Natural Resources). 2018. *Groundwater: Sustainable Groundwater Management Act.* Davis, Calif.: University of California DANR.

Di Matteo, L. and Dragoni, W. 2006. Climate Change and Water Resources in Limestone and Mountain Areas: The Case of Firenzuola Lake (Umbria, Italy). Proc. 8th Conf. on Limestone Hydrogeology, Neuchatel, Switzerland.

Dragoni, W. 1998. Some Considerations on Climatic Changes, Water Resources and Water Needs in the Italian Region South of the 43°N. *Water, Environment and Society in Times of Climatic Change* (Issar, A.S. and Brown, N., eds.). Dordrecht, the Netherlands: Kluwer. pp. 241–271.

Dragoni, W. and Sukhija, B.S. 2008. Climate Change and Groundwater: A Short Review. *Geological Society, London, Special Publications.* 288:1–12.

Gao, J. 2010. Collection and Analyses of Physical Data for Deep Injection Wells in Florida. A Thesis Submitted to the Faculty of The College of Computer Science and Engineering in Partial Fulfillment of the Requirements for the Degree of Master of Science, Florida Atlantic University, Boca Raton, Fla.

Hapgood, C.H. 1999. *Path of the Pole* (2nd ed.). Kempton, Ill.: Adventures Unlimited Press.

Heimlich, B.N.; Bloetscher, F.; Meeroff, D.M.; and Murley. J. 2009. *Southeast Florida's Resilient Water Resources: Adaptation to Sea Level Rise and Other Impacts of Climate Change.* Jupiter, Fla.: Center for Environmental Studies at Florida Atlantic University.

Herrera-Pantoja, M. and Hiscock, K.M. 2008. The Effects of Climate Change on Potential Groundwater Recharge in Great Britain. *Hydrological Processes.* 22:73–86.

Huntington, T.G. 2006. Evidence for Intensification of the Global Water Cycle: Review and Synthesis. *Journal of Hydrology.* 319:83–95.

IPCC (Intergovernmental Panel on Climate Change). 2007. Climate Change 2007: The Physical Science Basis. Contribution of Working Group I to the Fourth Assessment Report of the Intergovernmental Panel on Climate Change. Solomon, S.D.; Qin, M.; Manning, Z.; Chen, M.; Marquis, K.B.; Avery, M.T.; and H.L. Miller, H.L. (eds.). Cambridge, United Kingdom, and New York: Cambridge University Press.

Labat, D.G.; Goddéris, Y.; Probst, J.L.; and Guyot, J.L. 2004. Evidence for Global Runoff Increase Related to Climate Warming. *Advances in Water Resources.* 27:6:631–642.

Marshall, C.H.; Pielke, R.A.; Steyaert, L.T.; Willard, D.A. 2004. The Impact of Anthropogenic Land-Cover Change on the Florida Peninsula Sea Breezes and Warm Season Sensible Weather. *Monthly Weather Review.* 132:1:28–52.

Molak, V. 1997. *Fundamentals of Risk Analysis and Management.* Boca Raton, Fla.: Lewis Publishers.

Reilly, T.E.; Dennehy, K.F.; Alley, W.M.; and Cunningham, W.L. 2008. *Ground-Water Availability in the United States.* Reston, Va.: USGS (United States Geological Survey).

Scanlon, B.R.; Reedy, R.C.; Stonestrom, D.A.; Prudic, D.E.; and Dennehy, K.F. 2005. Impact of Land Use and Land Cover Change on Groundwater Recharge and Quality in the Southwestern US. *Global Change Biology.* 11:10:1577–1593.

Singer, S.F. (ed.). 2008. *Nature, Not Human Activity, Rules Climate Change: Summary for Policy Makers by the Nongovernmental International Panel on Climate Change.* Chicago, Ill.: The Heartland Institute.

Starkl, M. and Brunner, N. 2004. Feasibility Versus Sustainability in Urban Water Management. *Journal of Environmental Management.* 71:3:245–260.

Strzepek, K.M. and Yates, D.N. 1997. Climate Change Impacts on the Hydrologic Resources of Europe: A Simplified Continental Scale Analysis. *Climatic Change.* 36:79–92.

Thornthwaite, C.W. 1948. An Approach Toward a Rational Classification of Climate. *Geographical Review.* 38:1:55–94.

Wolock, D.M. 2007. *Estimated Mean Annual Natural Ground-water Recharge in the Conterminous United States.* Reston, Va.: USGS.

Chapter 3

Motivation for Groundwater Management

There are several reasons for utilities to assess and manage their water supplies. This manual opened with a discussion of the issue that caused the eastern Carolinas to change to a different water source. Fortunately, they had options. Having options is not always the case as noted in Chapter 2. The US Geological Survey (USGS) reports concerns with the Ogallala aquifer in the Great Plains; the Edwards aquifer in San Antonio, Tex.; the aquifers beneath South Dakota; groundwaters in eastern Wisconsin; and challenges throughout the desert Southwest to name a few. The concern is that many of these aquifers are overtaxed and their supplies unsustainable. Southwestern Utah (St. George / Cedar City) went through a similar exercise in the mid-2000s. The aquifer was dropping and the state geologist got involved. It was determined that groundwater was not available to supply the region. The region moved to a new impoundment meant to trap snowmelt for water supplies. The region will not be the last to seek alternative water supplies because of the lack of sustainable groundwater.

This chapter will focus on the quality issues associated with groundwater management. Keep in mind that upstream users can significantly impact the water quality that downstream users receive. The federal Superfund program is primarily about upstream contamination of groundwater and soils that can migrate and impact off-site users. The result is that groundwater contamination can take an apparent supply and make it unusable.

WATER QUALITY

The use of groundwater by utilities, agriculture, and industry is easy to understand. Small withdrawals are less likely to have major impacts on adjacent users of the aquifer; also, because costs are always a concern, groundwater is often perceived to be "better" than surface waters—or even "pristine," since contamination pathways into groundwaters are not as obvious as they are for surface waters. The perception is that less treatment will be required and well operations are straightforward. But this viewpoint is incorrect, as groundwater systems are as susceptible to contamination as surface waters and may be subject to contamination for more significant periods of time because of the slow movement of the water. Naturally occurring constituents, such as arsenic and barium, can impair groundwater quality.

Contamination includes metals, salt, arsenic, organics, and bacteria. Other contaminants are transported through the soils above the groundwater source from contaminant sources such as landfills and other hazardous waste sites and leaking underground storage tanks for chemicals and petroleum products. Pesticides and herbicides are transported off agricultural fields, along with phosphorous and ammonia fertilizers. Nitrates result from the breakdown of ammonia in fertilizer and septic tanks from bacteria. Unlined wastewater holding ponds and improper sludge or manure application sites are major contributors to nitrate contamination. Examples include *Escherichia coli* in a study conducted in Walkerton, Ont. (Schuster et al. 2005). Most users want to avoid these contaminants to the degree practical.

Groundwater-related disease outbreaks have risen as a result of contamination from biological contaminants, and other health conditions have increased as a result of contamination from chemical and biological agents. Of the 673 waterborne illness outbreaks that occurred in the United States between 1946 and 1980 (affecting 150,000 people), 44 percent were attributable to groundwater sources (Asano 1985). Furthermore, Keswick et al. (1982) estimated that 50 percent of the waterborne illnesses in the United States in a given year originate from groundwater, and 65 percent of those are enteric viruses (Yates et al. 1985). More recently, between 1997 and 2006, 137 waterborne disease outbreaks were reported to the Centers for Disease Control and Prevention, with 8,498 illnesses and 17 deaths (Yoder et al. 2008; Liang et al. 2006; Blackburn et al. 2004; Lee et al. 2002; Barwick et al. 2000). Of the 101 groundwater outbreaks with a known cause, 17 were attributed to chemical or toxin poisoning and 84 to pathogens. Bacteria were the most commonly implicated pathogen (Bloetscher et al. 2010).

In addition, many layers of rock that were previously thought to be highly confining may contain fractures or intrusions, such as wells that create pathways for contaminants to move into underlying aquifers. As a result, groundwater may also be "self-contaminating." Groundwater aquifers may be physically connected, which allows movement of lower-quality water into the production zone. Normally, this is caused by improper well design or construction, which allows poorer-quality water to migrate from one formation to another through potential protective or confining layers. Self-contamination can also occur if the water table lowers, causing saltwater to intrude from below or laterally from saltwater bodies. Saltwater intrusion is an example of the production zone being contaminated by adjacent saltwater bodies via upconing (pumping the wells too much, thereby pulling contaminants, typically salt, upward to the well), lowering regional groundwater levels (such is the case in South Florida) or overpumping along the coast.

The issue also manifests when wells are drilled deeper because the groundwater level drops. Drilling deeper may solve immediate issues with water quantity, but this decision may cause more problems, especially if the drilling occurs in the same exhausted aquifer. Deeper waters generally tend to have poorer water quality as a result of having often been in contact with the rock formation longer and dissolving the minerals in the rock into the water. Therefore, the types of rock strata that underlie the area have significant bearing on the quality and quantity of groundwater available and the ability of the rocks to store water.

While some deep aquifers may be prolific, the quality of water obtained from a well may not be desirable or even usable for drinking water without substantial amounts of treatment. Older water, and typically deeper water, poses greater challenges for treatment. The result is the potential for poorer water quality in the wells, along with greater aquifer drawdowns, accompanied by aquifer mining and land subsidence. Ideally, source water should be free of microbial contamination from human and animal activities and free of toxic natural and synthetic chemical contaminants; it should also be of adequate quantity to meet the needs of the population served. Having fewer contaminants means lower treatment costs as each increment of treatment to address constituents in the water increases the cost of the finished water. The real and unrealized costs to other users are often ignored in the current water management framework (Bloetscher and Muniz 2008). Since groundwater is rarely pristine and the potential for contamination is significant, water purveyors that derive their water sources from groundwater need to monitor surface events, such as rainfall, spills, development, and drought, to determine the potential impact on their water systems. That is where source water protection comes in.

SOURCE WATER PROTECTION

The 1996 amendments to the Safe Drinking Water Act require the states to assess areas serving as sources of drinking water to identify potential threats and initiate protection efforts. Most states delegated this requirement to local entities. Watershed protection can be broadly defined as a program to reduce the threat to water supplies from contaminants; for groundwater, this is called a wellhead protection program. Local communities protect their groundwater resources by incorporating wellhead protection activities into land use management as a means to reduce contamination. The protection of public water supply wells through wellhead protection programs is considered an important component of a comprehensive state groundwater protection program, as established by the US Environmental Protection Agency (USEPA 1991).

When a site has been selected for installation of a well or well field, the water supply entity must protect it from contamination through a wellhead protection regulation. A wellhead protection program reduces the potential for contaminants to leach into the groundwater by identifying and managing recharge areas specific to the wellfield. Wellhead protection programs range from simple regulations concerning the location of facilities in the vicinity of the wells to extensive land purchases and comprehensive land use restrictions. Typical action involving wellhead protection efforts include:

- new requirements for installation and testing of underground storage tanks;
- increased regulation for handling, using, and transporting toxic chemicals to reduce the possibility of spills;
- greatly increased regulation of landfills and other waste disposal sites;
- closer control of the use of pesticides and agricultural chemicals;
- sampling and monitoring of identified groundwater contamination locations; and
- action plans to remove the contamination.

Most states have groundwater programs. For example, California has established a process for developing groundwater management plans for individual aquifers. Connecticut ranks aquifers according to levels of protection. Texas has begun to develop regulations for aquifer management. The regulations are constantly in flux, so engineers and hydrogeologists need to contact the state's or province's division of water resources (or similar entity) to inquire about specific regulations for a given locale. AWWA's standard for source water protection ANSI/AWWA G300 provides a framework to protect source water.

Developing a Source Water Protection Program

Source protection is the first step in safeguarding public water supplies, reducing the need to use expensive alternative treatment techniques. Source water management options or tools can be broadly classified between regulatory and nonregulatory options. At the local level, the regulatory practices are usually in the form of zoning ordinances, subdivision ordinances, or regulations assigned directly to the protection of groundwater. The USEPA *Handbook for Ground Water and Wellhead Protection* (1994) indicates several options for implementation of a source water protection program including land use practices, regulations or legal measures, and administrative considerations. Other regulations, such as the Class V rule of 1999 to ban motor vehicle waste disposal wells and large-capacity cesspools, help protect aquifers.

The management of a source water protection area to prevent contamination involves the following:

- Identify protection options appropriate to the potential contaminants.
- Select the options that are technically and politically feasible.
- Implement the management practices.
- Develop contingency plans to address possible threats.
- Monitor the effectiveness of management practices.

To accomplish these goals, AWWA's G300, Source Water Protection, recommends the following steps:

- creating a vision,
- stakeholder engagement,
- characterization,
- goal setting,
- development of an action plan,
- implementation of action plan, and
- evaluation and revision as needed.

A groundwater-monitoring program is necessary to monitor the success of any source water protection program.

Delineation of a source water protection area is typically done through the use of computer modeling of travel time and pollutant transport. These models can be complex and can create large areas where many land uses are prohibited. Such source water protection efforts can conflict with private property rights objectives. These laws indicate that if the property is damaged significantly (often defined as a devaluation of 10 percent), the regulating agency must compensate the affected property owner. As a result, the concept of source water protection conflicts with this legal concept, which could be a significant impediment for utilities attempting to implement source water protection programs.

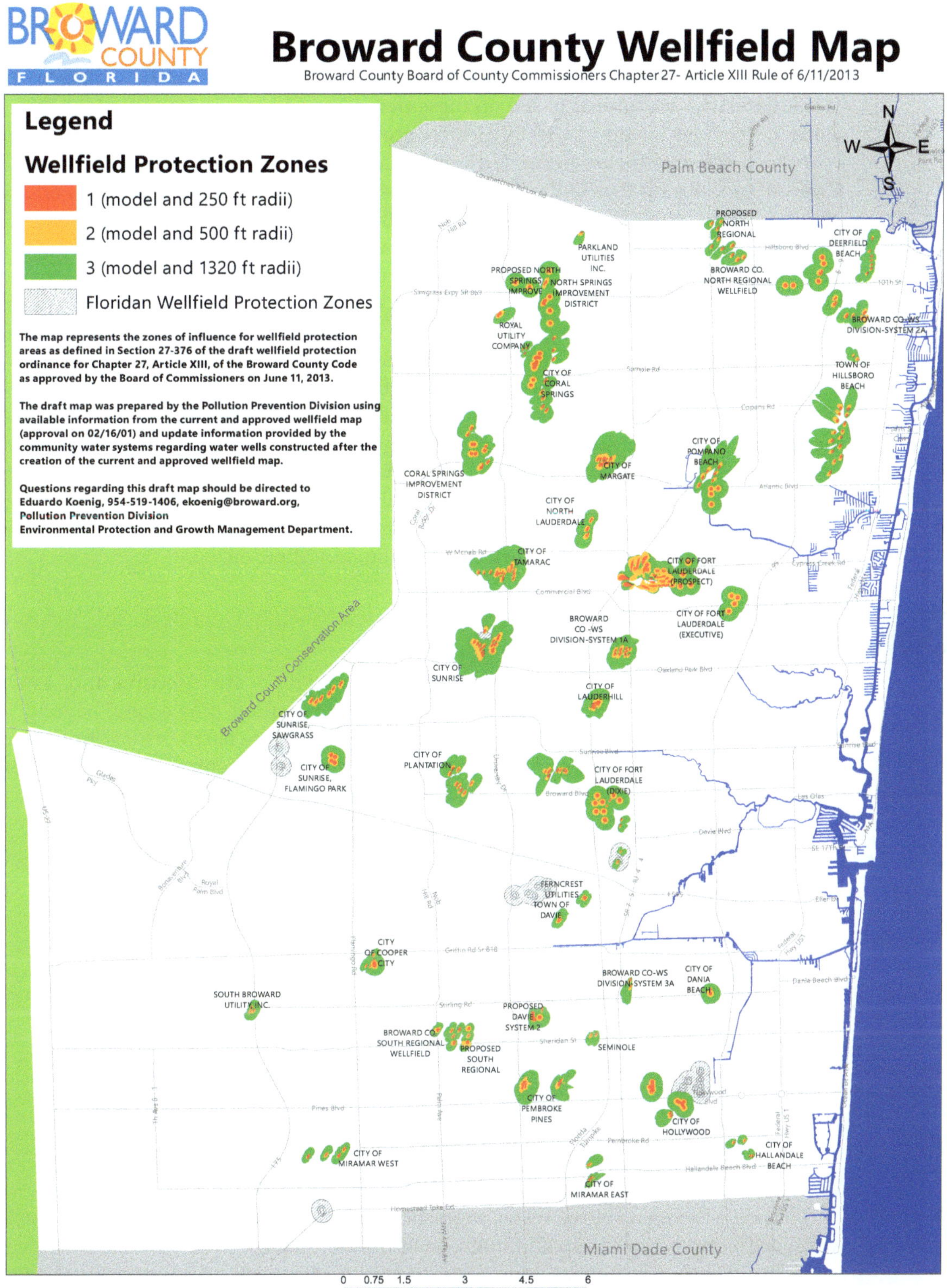

Source: Broward County, Fla.

Figure 3-1 Broward County source water protection zones

EXAMPLE OF A WELLHEAD PROTECTION ZONE

Figure 3-1 is an example of the source water protection areas from Broward County, Fla. The county has four tiers for regulation based on travel times from the surface to existing wells. Activities are limited based on these tiers (Tier 4 has no requirements):

Zone 1: Provides for up to a 10-day buffer around the well field. No hazardous chemicals (regulated substances) are permitted within Zone 1.

Zone 2: Provides up to a 30-day buffer. Businesses are required to be licensed and test the groundwater at their facility for regulated substances they store or use on site.

Zone 3: Provides up to a 210-day buffer. Businesses are required to be licensed, and secondary containment is mandated for their stored regulated substances.

In conjunction with source water protection efforts, water systems (municipal utilities, independently owned utilities, and other water management agencies) should identify any groundwater sources they are using that may be directly affected by surface waters. The concern is that if there is minimal filtration that occurs between the surface and the water withdrawn for wells, pollutants—especially microbiological constituents—may contaminate the water source. Waters that meet these criteria are considered "groundwater under the direct influence of surface water (GWUDI)." If an aquifer is determined to be GWUDI and, therefore, vulnerable to contamination by disease-causing organisms found in surface water, the well water must be treated under the same requirements as a surface water system, requiring mandatory disinfection and filtration.

GROUNDWATER PROTECTION STRATEGIES

The problem with groundwater protection is that aquifers are extensive and, as a result, often transcend local jurisdictional boundaries. Some aquifers may extend beyond state or provincial control. As a result, the varying institutions create barriers that offer formidable challenges to source water protection. For example, different entities may coexist in aquifers that:

- supply public municipal water,
- supply private users,
- deliver imported water,
- recharge the aquifer,
- control flooding,
- oversee quality,
- oversee quantity,
- manage irrigation supply,
- manage cleanup,
- manage biological habitats and species, and/or
- dispose of wastes.

In developing a source water protection program, water suppliers must involve the various diverse perspectives and interests of the community to be successful.

Source water protection requires cooperative efforts on all governmental levels and between units of government because of the movement of water across jurisdictional boundaries. To resolve water quality management issues associated with water supplies, agreements are normally developed (as arduous as this may be). The creation of agreements to alter the management, institute cleanup, or improve protection of a groundwater aquifer requires negotiation and a regional outlook by each entity.

CONTAMINATED GROUNDWATER MANAGEMENT STRATEGIES

Source water protection programs should be responsive to local needs and allow local autonomy, which is important for many governmental jurisdictions. Situations where local autonomy and interlocal cooperation may conflict identify where multijurisdictional cooperation is needed; the impact of contamination will impact local decisions. But what contaminants should utilities look for? More than 100,000 synthetic chemicals are in common commercial and industrial use in the United States—a number that continues to grow every year. Few are tested for, yet more than 200 of these chemical substances have been found in groundwater. Others may occur in groundwater where wells are not currently being drilled or investigation/testing has not occurred. Typically, concern does not arise until a contaminant is found in a potable supply. Organic chemicals, for example, have become a pervasive contaminant in groundwater supplies.

A serious water quality issue is perfluoroalkyl and polyfluoroalkyl substances. They have been in commercial use since 1940 because of their resistance to heat, water, and oil. Over the years, these materials have found their way into groundwater near airports, factories, and military bases. They are common in household cookware, lubricants, and the like. In the past several years, more than 50 bills in Congress were introduced. Two states have regulatory limits (California 5.1–6.5 ppt and New Hampshire 12–15 ppt). USEPA is looking at them as the next substances to regulate. The health concerns continue to be investigated. But as noted in Bloetscher and Plummer (2011), there are many potential others.

If a well becomes contaminated, the least-expensive and most-common strategy is blending the contaminated water with another source to meet drinking water standards, which does not solve the contamination issue but may resolve the regulatory concern. Water of differing types of contamination may be blended together to reduce their respective problems to acceptable levels. Sources include:

- a cleaner part of the aquifer,
- a different zone within the aquifer,
- another aquifer, or
- imported supplies.

In the pump-and-treat strategy, groundwater is extracted from an aquifer and treated. In some cases, the groundwater is treated beyond the required level and then blended with nontreated groundwater to produce a more cost-effective, potable supply. Treatment is generally more expensive than blending and may not always be technologically feasible. However, if the cost of water imported to supplement local supplies increases and its availability diminishes, the treatment of lower-quality groundwater can become feasible.

REFERENCES

Asano, T. 1985. *Artificial Recharge of Groundwater*. Stoneham, Mass.: Butterworth Publishers.

Barwick, R.S.; Levy, D.A.; Craun, G.F.; Beach, M.J.; and Calderon, R.L. 2000. Surveillance for Waterborne-Disease Outbreaks—United States, 1997–1998. *Morbidity and Mortality Weekly Report.* 49:SS04:1–35.

Blackburn, B.G.; Craun, G.F.; Yoder, J.S.; Hill, V.; Calderon, R.L.; Chen, N.; Lee, S.H.; Levy, D.A.; and Beach, M.J. 2004. Surveillance for Waterborne-Disease Outbreaks Associated with Drinking Water—United States, 2001–2002. *Morbidity and Mortality Weekly Report.* 53:SS08:23–45.

Bloetscher, F. and Plummer, J.D. 2011. Evaluating the Significance of Certain Pharmaceuticals and Emerging Pathogens in Raw Water Supplies. *Environmental Practice.* 13:3:198–215.

Bloetscher, F.; Meeroff, D.E..; Pisani, J.V.; and Long, S.C. 2010. Resolving Problematic Biofilms in Buildings and Compounds. *Journal of Environmental Engineering Science.* 27:9:767–776.

Bloetscher, F. and Muniz, A. 2008. Water Supply in South Florida—The New Limitations. Proc. AWWA 2008 Sustainable Water Sources Conf., Reno, Nev.

Keswick, B.H.; Gerba, C.P.; Secor, S.L.; and Cech, I. 1982. Survival of Enteric Viruses and Indicator Bacteria in Groundwater. *Journal of Environmental Science and Health.* A17:6:903–912.

Lee, S.H.; Levy, D.A.; Craun, G.F.; Beach, M.J.; and Calderon, R.L. 2002. Surveillance for Waterborne-Disease Outbreaks—United States, 1999–2000. *Morbidity and Mortality Weekly Report.* 51:SS08:1–28.

Liang, J.L.; Dziuban, E.J.; Craun, G.F.; Hill, V.; Moore, M.R.; Gelting, R.J.; Calderon, R.L.; Beach, M.J.; and Roy, S.L. 2006. Surveillance for Waterborne Disease and Outbreaks Associated with Drinking Water and Water Not Intended for Drinking—United States, 2003–2004. *Morbidity and Mortality Weekly Report.* 55:SS12:31–58.

Schuster, C.J.; Ellis, A.G.; Robertson, W.J.; Charron, D.F.; Aramini, J.J.; Marshall, B.J.; and Medeiros, D.T. 2005. Infectious Disease Outbreaks Related to Drinking Water in Canada, 1974–2001. *Canadian Journal of Public Health.* 96:4:254–258.

USEPA. 1994. *Handbook for Ground Water and Wellhead Protection.* Washington: USEPA.

USEPA (US Environmental Protection Agency). 1991. *Protecting the Nation's Groundwater: EPA's Strategy for the 1990s.* Washington: USEPA.

Yates, M.V.; Gerba, C.P.; and Kelley, L.M. 1985. Virus Persistence in Groundwater. *Applied and Environmental Microbiology.* 49:4:778–781.

Yoder, J.; Roberts, V.; Craun, G.F.; Hill, V.; Hicks, L.; Radke, V.; Calderon, R.L.; Hlavsa, M.C.; Beach, M.J.; Roy, S.L. 2008. Surveillance for Waterborne Disease and Outbreaks Associated with Drinking Water and Water Not Intended for Drinking—United States, 2005–2006. *Morbidity and Mortality Weekly Report.* 57:SS09:39–62.

Chapter 4

Groundwater/Aquifer Characterization Tools

Given water supply and quality challenges, groundwater management requires information on the aquifer itself. This chapter provides an overview of the technical information needed to adequately describe, or characterize, a groundwater resource intended for use by a water utility, including those parameters of the aquifer needed to create a useful management plan and model of the aquifer system. These are fundamental tools accessible to water utilities and include physical, geological, and water quality parameters. Topics and tools address:

- data needs for developing a groundwater budget,
- parameters and methods for delineating the aquifer,
- fundamental concepts controlling well and aquifer yield, and
- introductory information regarding groundwater quality.

GROUNDWATER BUDGET

A groundwater budget is simply an accounting of the inputs and outputs of a particular groundwater resource being considered for use for water supply purposes. For a groundwater resource from an aquifer to be considered sustainable, the water that is removed from the aquifer (discharge) must be replaced by water that is added to the aquifer (recharge). While the concept of water balance and groundwater budget is relatively simple, no one can indefinitely withdraw groundwater from an aquifer at a rate faster than it is recharged, but as noted in Chapter 2, this happens repeatedly. The possible reasons for withdrawals at rates greater than recharge include demands for water, the inability to fully measure the changes in the aquifer, poor testing, and failure to consider competing users and the specific components of water balance, which can be difficult to quantify. One or more can play a part in decisions. As a result, groundwater budgets must be considered as a part of overall water budgets for those utilities that rely on, or project use of, groundwater as a water supply.

Water availability is directly related to water balance. For a groundwater resource from an aquifer to be considered sustainable, the water that is removed from the aquifer (by pumping, discharge from springs, plant intake, and evaporation) must be replaced by water that is added to the aquifer. If the groundwater withdrawals from an aquifer exceed groundwater recharge, there will be no water availability at some point in the future. This is the fundamental driver for the need for the sustainable management of groundwater.

Groundwater discharges are divided into two categories: natural discharges and artificial (pumping) withdrawals. Human withdrawals are far more readily quantified because they can be easily determined with flow meters. Natural discharges are not metered and may be both site-specific and variable. Extensive efforts are required to determine the differences between natural and man-made discharge, creating uncertainty in the estimates. Recharge can be estimated with long-term, multi-seasonal aquifer testing, but not by short-term tests. Aquifer tests are estimates, and both localized differences and a tendency to overestimate create common errors.

AQUIFER CHARACTERIZATION

Where there is development of an aquifer for water use, there is usually a cadre of published reports about the aquifer. For example, the US Geological Survey (USGS) has published reports for decades on major aquifers across the United States with data and other information (see the USGS Publications Warehouse). In undeveloped regions, reconnaissance-type interpretation of aerial photos and maps may be the first step. In many regions, general cross sections and hydrogeologic interpretations will already exist, along with well logs and production information from existing wells. USGS creates many of these reports, but oil companies may have them as well as a result of previous oil and gas exploration. In such cases, cross-sectional maps may have to be drawn from surface geologic information and existing well logs. These maps will aid in determining the location, depth, and thickness of favorable aquifers.

In cases where favorable aquifers will have already been identified, exploration efforts can fill in the details of the local hydrogeologic environment. Surface geophysical methods, borehole geophysical methods, aquifer testing, exploratory drilling, and lithologic logging are the most common methods used to accomplish these efforts. More details on these are available in Bloetscher et al. (2008) or Keys (1989).

Delineation – Horizontal and Vertical Extent

An aquifer is not infinite. Therefore, finding the boundaries—changes in formations, water bodies, the ocean, etc.—will impact how the aquifer is modeled and how productive it will be. Reports, data, and information about regional and specific aquifers by USGS and others may be of help. There are two boundary conditions that are relevant (both in management and modeling): impermeable and recharge boundaries.

The effect of an impermeable barrier around an aquifer is that the aquifer is cut off on one side, which causes the effects of drawdown to multiply. An example is an impermeable barrier caused by the rising side of a buried valley, a fault that cuts off a formation, or a clay layer that bounds the aquifer on one side. This situation is quite common in once-glaciated parts of the United States.

A "recharge boundary" is created by the effects of a local stream or lake. When the recharge boundary is a water body, the aquifer is cut off on one side by a water body that recharges the aquifer. This is a situation that is often found in the field.

Aquifer and WellField Yield

Data referred to as aquifer parameters are needed to determine the well yield. The most significant aquifer parameters are porosity, transmissivity/hydraulic conductivity, specific yield and specific retention, hydraulic head, and gradient. *Porosity*, *specific yield*, and *specific retention* describe the rock formation and quantities of water existing in the formation. *Head* (pressure or vertical datum) and *gradient* determine how water moves through the formation and represent the mechanics of horizontal and vertical recharge to a well being pumped. Head and gradient are also used to analyze the transport of pollutants that may migrate to a well. *Transmissivity/hydraulic conductivity* indicates how easily water will move in the formation. These parameters are detailed in Appendix A as well as AWWA (2014).

Hydrologists divide groundwater into the portion that will drain under the influence of gravity, which is called *specific yield*, and the portion that is retained as a film on rock surfaces and in very small openings, which is called *specific retention*. The physical forces that control specific retention are the same forces controlling the thickness and moisture content of the capillary fringe.

Head is the elevation of the water in an unconfined aquifer or the pressure of a confined aquifer (see Figure 4-1). The depth to the water table affects the development of water supplies from unconfined aquifers. Where the water table is shallow, the land

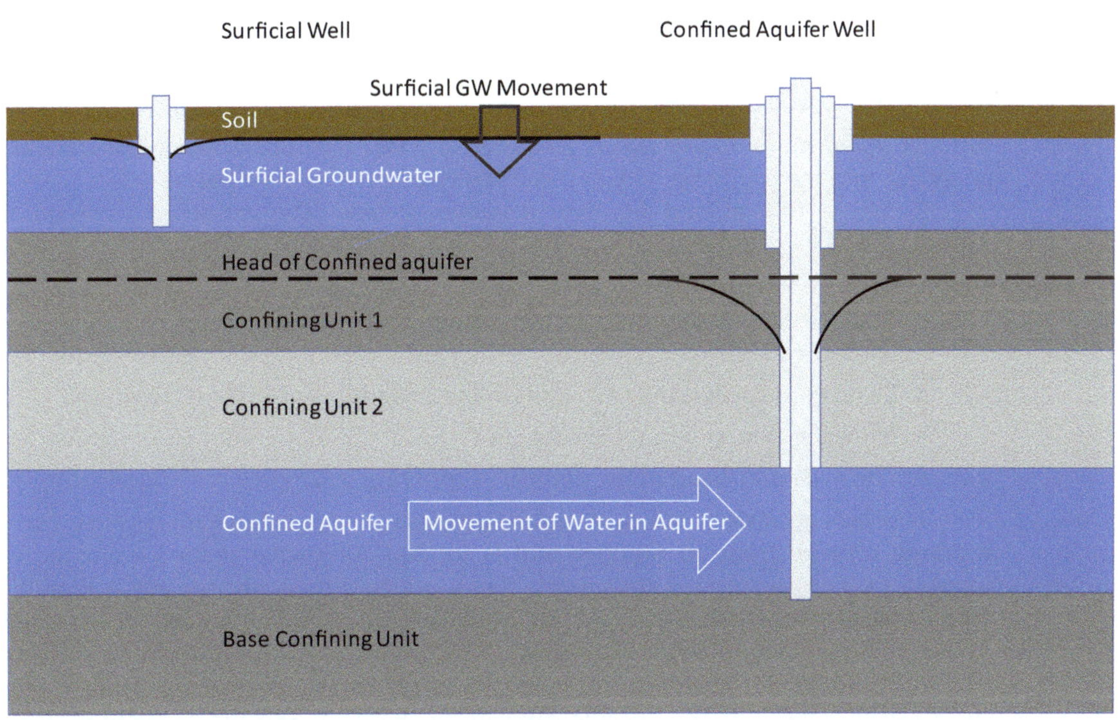

Source: Created by author

Figure 4-1 Diagram of wells and subsurface flow

may become waterlogged during wet weather and unsuitable for most land uses (such as housing or agriculture). Where the water table is at great depth, the cost of constructing wells and pumping water for domestic needs may be prohibitively expensive.

The depth to water in a nonflowing well is subtracted from the elevation of the measuring point to determine the total head at the well. Total head, as defined in fluid mechanics, is the sum of elevation head, pressure head, and velocity head. Because groundwater moves relatively slowly, velocity head can be ignored. Therefore, the total head at an observation well involves only elevation head and pressure head. Groundwater moves in the direction of decreasing total head (i.e., downhill), which may or may not be in the direction of decreasing pressure head.

All other factors being constant, the rate of groundwater movement depends on the *hydraulic gradient*. The hydraulic gradient is the change in head per unit of distance in a given direction. If the direction is not specified, it is in the direction in which the maximum rate of decrease in head occurs.

Hydraulic conductivity may also vary by direction at any place in an aquifer. If the hydraulic conductivity is essentially the same in all directions, the aquifer is *isotropic*. If it varies by direction, such as differences between conductivity in the vertical and horizontal directions, the aquifer is *anisotropic*.

Although it is convenient in many mathematical analyses of groundwater flow to assume that aquifers are both *homogeneous* and *isotropic*, such aquifers are rare, if they exist at all. Hydraulic conductivity in most rocks, especially in unconsolidated deposits and in flat-lying consolidated sedimentary rocks, is larger in the horizontal direction than in the vertical direction, usually by a factor of about 10.

Field Testing

Determining *transmissivity* or the storage coefficient by any means other than actual performance tests in the field is expensive, time-consuming, and of questionable accuracy. In confined aquifers, the testing notoriously overestimates available water, in part because recharge is often overestimated.

Field-testing methods for determining these values have been developed and are thoroughly documented. These methods apply a regulated stress (pumping) to the formation and measure the effects (changes in water level) produced. The data are then analyzed, and the transmissivity and storage coefficient are calculated.

Recharge to confined aquifers assumes *leakage*. The modified non-equilibrium formula for leaky artesian conditions is based on the conditions for confined aquifers and on the following assumptions:

- The aquifer is confined between an impermeable bed and a bed through which leakage can occur.
- Leakage is vertical into the aquifer and proportional to the drawdown.
- No water is stored in the confining bed.
- The hydraulic head in the deposits supplying leakage remains constant.

The leakage associated with many aquifers is overestimated because the actual recharge may be from an outcrop that is located somewhat distant from the area of interest. One of the important concepts that a utility should understand is that confined aquifer recharge can be restricted if the outcrop is developed. Outcrop development can also introduce contaminants upstream of the utility's water supply well.

An abbreviated well-performance evaluation can be performed using a relatively short test to determine the specific capacity of the well. Specific capacity is defined as the *yield* of the well, usually expressed in gallons per minute per foot of drawdown (gpm/ft [m^3/m*day]). The total *drawdown* in a well can be divided into two components—

drawdown in the aquifer and drawdown related to well loss. Drawdown in the aquifer depends on the aquifer's ability to transmit water, and it generally does not change unless the aquifer is being depleted. Drawdown because of well loss is related to the ability of the well to transmit water, and it may change with time because of turbulent flow or head loss as the water passes through the screen or well bore. Some of the factors affecting well loss include changes in the chemical or bacterial quality of the water or changes in the mechanical condition of the well itself.

An *unconfined*, or water-table, aquifer does not have water confined under pressure beneath impermeable rocks. Water is derived from storage by gravity drainage of the interstices above the cone of depression, by compaction of the aquifer, and by expansion of water in the aquifer.

Properties of an unconfined aquifer can be determined by the Theis equation under some limiting conditions. The Theis equation (Theis, 1935) describes radial confined groundwater flow in a uniformly thick horizontal, homogeneous, isotropic aquifer of infinite areal extent as follows (noting that most aquifers have none of these properties).

$$S = (Q / 4\pi T)\int\left(r^2 s/4Tt\right)\infty \left[(\text{e-u/u})\ \text{du}\right] \qquad \text{(Eq 4-1)}$$

Because the Theis equation is measuring the cone of influence of the well (which is assumed to be large compared to the radius of the well), the radius of the pumped well is assumed to be negligible. In Eq 4-1, Q is the discharge (L^3/T), T is the transmissivity (ft^2/time), t is the time interval of interest, S is the dimensionless storage coefficient, and r is the radial observation distance from the pumped well. The derivation and solution are documented in many places (Jacob 1940) and will not be discussed further here. One of the basic assumptions of the Theis equation is that water is released from storage instantaneously with a decline in head. In a water-table aquifer, this is not always true because water is derived partly from gravity drainage, and the effects of gravity drainage are not considered in the Theis formula. However, with long pumping periods, the effects of gravity drainage become negligible so that the Theis solution can be used.

Usually, the Theis equation is fitted graphically to experimental pump test data to obtain approximations for the storage coefficient (S) and the transmissivity (T). Because of these assumptions and the pump tests are usually of relatively short duration (a few hours to a few days), the equation should be assumed to incorporate significant uncertainty, but this is rarely the case.

EVALUATION OF CURRENT CONDITIONS

Two steps are used in identifying potential groundwater sources for a public water supply. The first is identifying regions having low pollution potential, high recharge capability, and a favorable location for the utility and its customers. The second step is performing field investigations to confirm site-specific characteristics of the aquifer system to determine how much water is potentially available. Care should be taken with confined aquifers where water withdrawal capability is consistently overestimated.

Additional items needed for the evaluation of regional groundwater conditions and the potential for resource development should be based on the following factors:

- long-term land uses affecting quality and quantity of groundwater recharge;
- legal issues such as water rights, adjudications, water transfers, and conjunctive use commitments;
- permits; and
- wellhead protection.

Land Use

Land uses—current, past, and future—may impact the siting of wells. As noted in the previous chapters, surface activities have the potential to affect groundwater quality. Potential development and impacts of that development should also be considered so that future land use does not degrade the water that flows off the land in the vicinity of the well. As land uses change, an aquifer with good quality at the time of development may deteriorate. Consequently, water resource professionals must obtain and maintain a good understanding of urban and industrial growth and zoning of the area associated with the groundwater supply. Testing for contaminants throughout a proposed well field is required by federal, state, and local regulations. Pollution source areas, especially within public water supply well capture zones and those that are currently upgradient of the proposed water supply site, are of utmost concern.

Today, simulation models (see Chapter 6) can depict the long-term effects of a proposed water-well development and simulate the potential impacts of any land-use changes that may affect the local hydrology. Examples are industrial development displacing irrigated agriculture or paving from urbanization that reduces recharge. Because of a lack of certainty about flow patterns and possible changes in flow patterns, judgment is required for all methods of groundwater modeling. In addition, flow of discharged material in the unsaturated zone is not always governed by groundwater gradients, and *dense non-aqueous phase liquid* (DNAPL) and *light non-aqueous phase liquid* (LNAPL) contaminants can move contrary to prevailing groundwater flow. Groundwater development should be upgradient or downgradient at an appropriate distance from potential threats to the water quality. Testing for pollutants at greater frequency may be appropriate. Establishing early-warning monitoring wells at various depths may be required to monitor groundwater quality.

Existing Water Rights

The development of new wellfields should be shown to not infringe on existing water rights or competing water users. The water developer should be able to demonstrate "sustainable yield" and also show that the side effects of pumping, such as land subsidence or saltwater intrusion, or interference with competing water or competing land users would be negligible. Often, this can be accomplished via computer modeling by using a program such as USGS's MODFLOW program (see Chapter 6). These models can also be used to demonstrate to regulatory agencies that proposed drawdowns will not be detrimental to the environment or adjacent users, or exceed the drawdowns set by regulation. More importantly, these models can be used to identify where less pumping and better groundwater management practices should be employed. Aside from the investigation of the hydrogeology, water quality should also be considered. Computer programs that include water quality changes with time are also available for use with groundwater flow models.

Changes Affecting Evaluation

Changes in economics may influence local officials to develop more groundwater resources or the public to demand higher-quality water. Industry and development changes may be at odds with the utility provider's aquifer protection programs or make public relations efforts difficult. Reliable measurements must be obtained when evaluating any groundwater system.

Deterioration of water quantity and quality can have grave consequences for those dependent on the water resource in the form of higher treatment costs or abandonment of the well field. However, changes other than those measured in the field can often be of great concern. Perceived changes may be caused by the availability of improved analytical methodologies (which detect levels of compounds not previously measurable or of new chemicals); regulatory priority shifts toward or away from protection of natural resources, including aquifer classifications; new toxicological data or reinterpretation of existing toxicology; or the integration of facts regarding groundwater quality.

Exploratory Drilling

In some instances, existing wells may not be located in a potential water supply aquifer, and drilling one or more exploratory wells into the aquifer is necessary. Exploratory drilling is performed to determine an aquifer's characteristics, including hydraulic conductivity, water quality, thickness, and areal extent. The newly drilled wells can also be used for borehole geophysical logging and aquifer testing. The number and spacing of the wells drilled will depend on the size of the aquifer and the amount of water supply development planned.

Borehole Geophysical Logging

Borehole geophysical logging has become a common tool in groundwater exploration. The technology of borehole logging is quite involved, and experts using special equipment are needed to perform the logging and interpretation. There are many texts on borehole logging, but the reader should be cautioned that most of these texts present information tailored for use in petroleum exploration. The presence of low- or non-saline water in a formation mandates the use of special analyses for interpretation. For a detailed discussion of this subject, see *Borehole Geophysics Applied to Groundwater Investigations* (Keys 1989) or Keys and MacCary (1971).

Aquifer Testing

In the simplest form, recording flow rates of the water produced at different depths while drilling with air is a form of aquifer testing, yielding valuable information. Aquifer testing is useful during exploration of test holes, and many effective methods are now available for performing such testing. Methods of aquifer testing are discussed in detail in Bloetscher et al. (2008) and many groundwater texts.

Land Costs

Legal access to property for preliminary groundwater investigation will commonly be granted by the landowners on request. However, as soon as a property has been identified as attractive for detailed exploration, purchase or lease options should be obtained. The cost of land may be more reasonable before good water supplies are confirmed. Drilling-site preparation and restoration can be added costs to the groundwater development that should not be overlooked.

Permits

Exploratory drilling permits must be secured and fees paid, often at the state or county level. Obtaining permits is usually the responsibility of the well driller or the engineer in charge. In addition to the driller, a qualified consulting hydrogeologist or engineer often supervises the drilling and well-construction activities. The consultant's responsibilities will usually include identifying appropriate well-construction materials, reviewing well logging data, conducting or overseeing geophysical logging, interpreting logs, well design, and certifying as-built drawings.

Part of the drilling permit will include proper abandonment once the well has served its purpose. Regulatory agencies are concerned about well construction and post-use of a well because of the possibility of cross-contamination between shallow zones and deeper high-quality aquifers. The documentation of all fieldwork can be valuable in later phases of groundwater development or protection. Proper land survey location, global positioning system (GPS) locations, description of the wells, and complete as-built drawings of construction are desirable.

Professional Services

Consulting firms specializing in hydrogeology are valuable to the utility for their knowledge and experience in carrying out fieldwork and preparing the necessary reports. Some firms offer total services in groundwater development, as well as highly specialized equipment or services. Budget constraints, the complexity of the project, and the adequacy of the utility's staff can determine the most appropriate team. Some states require registered engineers, certified geologists, and other professionals to verify the accuracy and completeness of fieldwork.

REFERENCES

AWWA. 2014 (4th ed.). M21, Groundwater. Denver, Colo.: AWWA.

Bloetscher, F.; Muniz, A.; and Largey, J. 2008. *Siting, Drilling and Constructing Wells.* Denver, Colo.: AWWA.

Jacob, C.E. 1940. The Flow of Water in an Elastic Artesian Aquifer. *Transactions of the American Geophysical Union.* 21:2:574–586.

Keys, W.S. 1989. *Borehole Geophysics Applied to Ground-water Investigations*. Westerville, Ohio: National Water Well Association.

Keys, W.S. and MacCary, L.M. 1971. Application of Borehole Geophysics to Water-Resources Investigations. *Techniques of Water-Resources Investigations*, Book 2, Chap. E1. US Geological Survey.

Theis, C.V. 1935. The Relation Between the Lowering of the Piezometric Surface and the Rate and Duration of Discharge of a Well Using Groundwater Storage. *Transactions of the American Geophysical Union.* 16:2:519–524.

Chapter 5

Monitoring to Support Management of Regional Groundwater Systems

To manage groundwater, a monitoring program needs to be in place. The monitoring program permits the utility to collect data and use these data for trend analysis and for computer modeling if that is desired. Modeling the aquifer using computers can help with the analysis and understanding of the reasons behind long-term trends and the potential risks that arise, whether from pollutants, upconing, or incorrect assumptions with respect to recharge. The objectives of and protocols for a groundwater monitoring program are typically documented in a Groundwater Monitoring Plan. The goal of a Groundwater Monitoring Plan is to document and stabilize how water-related data are collected, evaluated, analyzed, and used in making water management decisions. Often accompanying a monitoring plan are a Sampling and Analysis Plan (SAP) and a Quality Assurance Project Plan (QAPP). Many documents describing the preparation and use of SAPs and QAPPs are available on the internet. See the US Environmental Protection Agency (USEPA) website for references relating to the development and implementation of these documents.

Typical components of a Groundwater Monitoring Plan include:

- description of the monitoring plan objectives and rationale, including a history of groundwater monitoring in the study area;
- description of the role of cooperating agencies (if applicable);

- description of the physical, hydrologic, and hydrogeologic characteristics of the study area to be monitored. This may include a discussion of the aquifer, along with typical weather patterns and surface water bodies in the study area;
- discussion of any monitoring network, including a map showing the well locations, and a discussion regarding the rationale for the well selection;
- construction and development details for wells included in the monitoring network;
- monitoring schedule;
- description of field methods/protocols for collecting data and/or water samples, including chemical analysis protocols for analytical samples;
- data evaluation and analysis methods; and
- reporting protocols.

Monitoring plans should be considered dynamic documents and should be revisited on a regular basis. Similarly, monitoring networks and their resultant data sets should be evaluated on a regular basis to identify data gaps and to plan for the collection of data to address those gaps.

MONITORING WELLS AND NETWORKS

A monitoring network is a series of groundwater wells, surface water gauges, and meteorological stations from which data are collected, recorded, and analyzed regularly. The purpose of the network is to provide data that, when evaluated as a whole, provide a current picture of hydrologic and hydrogeologic conditions in the study area. A robust monitoring network is required to address long-term reliability and sustainability of the aquifer and for delivery of water supply to the utility's customers.

A groundwater monitoring network should contain a sufficient number of wells to evaluate and characterize a groundwater basin both spatially across the basin and with sufficient different depths such that each separate water-bearing zone is monitored. The typical goal of the monitoring program is to identify and/or confirm groundwater trends, provide data for the basin conceptual model, and/or fill specific data gaps. Groundwater elevation and quality data should be combined with surface water data, meteorological data, and other data such as subsidence figures to create the necessary conceptual model of the basin. Monitoring networks can be designed by using statistical methods for optimizing well placement or for designing the monitoring program to be implemented. By thinking about how the data from the network will be evaluated, it becomes easier to develop the actual network design.

Data from monitoring networks should represent each individual aquifer/water-bearing zone in the basin and should be able to evaluate short-term, seasonal, and long-term trends in basin conditions. Data from these wells (and, therefore, the number, location, and depth of monitoring wells) should be sufficient for mapping groundwater depressions, recharge areas, and along margins of basins where groundwater flow is known to enter or leave the basin. As with monitoring plans, monitoring networks should be assessed at least every five years and improvements/modifications made if necessary.

Ideally, wells contained in a monitoring network are dedicated groundwater monitoring wells with known construction information that meets applicable well standards set by individual states or professional organizations. Other wells may be used in a network, but data from those wells should be considered and used with caution. Most importantly, selected monitoring wells should not be influenced by factors related to other wells in the vicinity (e.g., pumping) as this may result in erroneous data. Additionally, activities that directly or indirectly affect the quantity and quality of groundwater in the study area must be considered in both the design of the network, to minimize these impacts, and in the use of the data from the network. In a nutshell, optimization of a monitoring network requires the wells and monitoring points within the network to be well thought out and organized with respect to the number and location of wells and the depths from which data are collected (relative to the overall hydrogeologic understanding of the basin).

In areas where contamination risks are high, drawdowns are increasing, or declining water levels may indicate mining, sentinel monitoring wells should be installed. These wells, located at various depths, will provide definition for the initial groundwater assessment. Sentinel wells also serve as an early-warning system to detect changes in water quality and water elevations before they affect the water supply wells. In recent years, equipment for sampling monitoring wells has become widely available. Small submersible or portable pumps can be installed into well casings as small as 1 in. (25 mm) in diameter, although it is common for monitoring wells to be at least 4 in. (100 mm) in diameter and made with polyvinyl chloride (PVC) casings.

The number of wells needed and their locations, depths of completion, and construction details must be specified as part of an integrated plan. The plan must account for likely sources of contamination, local hydrogeology, and the hydraulic effects of the proposed groundwater development. For example, what was previously considered downgradient from the well can become upgradient either after pumping begins or as influenced by nearby surface water. Therefore, monitoring wells must be installed upstream as well as downstream of the proposed water supply well in the event the well pumping alters the aquifer flow regime. These changes should be simulated with computer modeling to aid in designing a monitoring-well network. The initial assessment may indicate that one monitoring well is sufficient to begin with, but an increased contamination threat in future years (e.g., because of local growth and development) could indicate a need for additional monitoring wells. This potential for development demonstrates the importance of continued evaluation of changes that might affect the groundwater supply.

Defining a Strategy for Monitoring Needs

In establishing a monitoring network and developing a monitoring plan, the goals of the monitoring program must first be identified and a strategy developed to identify monitoring needs. Questions to be considered include:

- What types of data are needed (i.e., groundwater elevation; vadose zone data; extensometers [for land subsidence]; surface water elevations in streams, rivers, and lakes)?
- What existing data are there? What is the quality and quantity of those data?
- What is the minimum level of data requirements?
- What questions need to be answered? As a result, how accurate do the data measurements need to be?
- How good does the quality need to be?
- How often should data be collected?
- What trends need to be discerned (annual, seasonal)?

Once the utility figures out what they are trying to achieve with the monitoring program (and what questions they are trying to answer) and takes a detailed look at data that are already available (including the spatial and temporal variations in those data, as well as the data quality), they need to next identify what is not known. Some data gaps will be apparent, and others will only become evident when actively trying to answer a specific question (e.g., At what groundwater elevations does inelastic subsidence begin in the groundwater basin?). After establishing the data gaps, a plan will need to be developed to fill those gaps. In monitoring, there is no such thing as one-stop shopping; filling data gaps takes time and will need to be implemented over the long term. Remember, several questions that need to be answered via monitoring involve the temporal variability of data.

If the utility is trying to assess a well's vulnerability to contamination, the utility staff would want to be familiar with the information in the US Geological Survey (USGS) Circular 1385. It can be used to devise improved monitoring programs to identify the most beneficial protection programs for a particular well.

Monitoring Well Installation

Most states have well construction standards that should be followed in well design and construction; see ASTM/AWWA A100, M21, Bloetscher et al. (2008), and various ASTM standards such as ASTM D5092 / D5092M Standard Practice and Installation of Groundwater Monitoring Wells. The reader is encouraged to consult these references.

Monitoring wells are designed and located to provide representative groundwater data and samples to assess groundwater levels and quality over time. Monitoring wells can also be used to characterize subsurface lithology (during construction); obtain hydraulic head information to estimate groundwater flow direction, gradients, and velocities; conduct aquifer pumping tests to estimate aquifer hydraulic properties (such as transmissivity); and provide estimates of mass flux and discharge (a lot of the information discussed on Chapter 4). If the monitoring wells are installed to track a specific contaminant, the type of constituent must be considered so that the well can be screened appropriately and designed to prevent cross-contamination to unimpacted areas of the aquifer system.

As a general rule of thumb, all monitoring wells should prevent the infiltration of surface water into a well, prevent cross-contamination within or between water-bearing zones, and be secured at the surface to prevent unauthorized access. In addition:

- Monitoring wells should be designed with respect to the well's function within a well network.
- Wells and borings should be logged during drilling by someone experienced in hydrogeology.
- Well names and numbers should be permanently assigned and never change. The approach to naming and numbering of wells should be carefully considered—sequential names and numbers are often used and area easily understood. Replacement wells often have an "R" added to the well name to indicate that they are replacement wells.

There are different types of monitoring wells:

- Traditional (vertical) wells with one single interval of well screen.
- Traditional (vertical) wells with multiple intervals of well screen – same as above though there are annular seals placed between filter pack sections adjacent to well screen intervals.
- Nested wells – multiple vertical wells constructed in a single borehole. Nested wells are used to monitoring multiple groundwater zones simultaneously. They allow for evaluation of vertical movement of groundwater. Another advantage is that they take up less space than multiple wells but are tricky to install because of the need for annular space around well (which requires a large-diameter borehole). There are commercially available multi-level monitoring systems. The costs of these systems are typically greater than the cost for conventional monitoring approaches but can provide better quality installation.
- Clustered wells – multiple traditional wells with varying screen depths installed in a same general area (each well in its own borehole). These wells obtain the same

benefits of nested wells without concerns about construction, but they take up more space and can cost more because of multiple mobilizations of the drilling rig.

- Direct push (DP) / pre-packed wells – similar in overall design and construction to typical groundwater monitoring wells, except that the diameters of the borehole and well casing are generally narrower, pre-paced well screens may be used, and the well installation method is different. DP wells are small-diameter hollow steel rods or PVC pipes that are pushed, driven, or vibrated into the ground. They can be installed in a cluster or as included wells and can be used as piezometers. DP wells installed in unconsolidated sediments can often be advanced to depths of around 100 ft. They are not suitable for coarse, consolidated, cemented, or lithified deposits or in bedrock. Care must be taken to ensure an adequate bentonite seal and annular seal during installation. Bentonite sleeves may reduce the risk of vertical cross-contamination. The biggest advantage is that they are fast and relatively inexpensive to install and generate very little waste (cuttings).
- Angled (or inclined) wells – use specialized drilling rigs to advance angled borehole. Well construction in these boreholes is tricky as you want to get an even filter pack in the annular space around the well.

Traditional (vertical) monitoring wells are designed considering:

- drilling method,
- borehole diameter,
- annular space,
- borehole alignment,
- total depth of the borehole,
- selection of annular materials,
- well development, and
- workplace safety.

The depth of monitoring wells is determined by site-specific hydrogeologic conditions and purpose. Regardless of monitoring depth, the depth of completion of a monitoring well in a borehole should be within 1 ft of the bottom of the screen interval. If the borehole is greater than the final well depth, the borehole should be backfilled with low-permeability material (e.g., bentonite grout) to the final depth of the well.

Monitoring Protocols

There are many references available from national and state organizations providing detailed information on standard operating protocols or procedures for monitoring. Rather than duplicate the information provided in these documents, the following is a discussion of general protocols to be considered as part of a groundwater basin monitoring program.

First, establish the monitoring sites. Make sure that there is long-term, year-round access, and complete any necessary agreements with private landowners to access the monitoring wells, staff gauges, or other monitoring points. Next, establish unique identifiers for each monitoring location and document those along with a written description of the monitoring point, the date established, a point of contact (for access if needed), the type of monitoring (e.g., well, staff gage, lysimeter), the location of monitoring point (latitude and longitude), the elevation, etc. Any modifications made to the monitoring location should be tracked in a permanent log.

When installing wells for groundwater elevation or quality monitoring, the borehole must be logged under the supervision of a professional geologist, professional engineer, or other professional experienced with drilling and identifying hydrogeologic strata.

A written criteria for logging of borehole cuttings and conducting borehole geophysical surveys (resistivity, spontaneous potential, spectral gamma, or other as appropriate) should be developed to confirm the hydrogeologic logs. Caliper logging should also be performed. Finally, if appropriate, a well completion report should be prepared and submitted to the appropriate regulatory agency(ies), including logs as an attachment.

Protocols for measuring groundwater level and quality generally require that data are taken from the correct location and are documented in a field log, including well ID, screened interval, and depth. If possible, data should be collected multiple times in the same location to ensure that the data are accurate and reproducible. Data to be collected should represent conditions that are useful to inform the utility about the appropriate basin management decisions that need to be made. Finally, all salient information should be recorded to normalize the data (if necessary), compared to prior data, and handled in such a way as to preserve data integrity.

Generally, well-monitoring information to be collected and logged includes groundwater elevations that will be used to form the basis of basin-wide piezometric maps and should approximate conditions at a specific point in time. For groundwater elevation monitoring, depth to groundwater must be measured relative to an established reference point (RP) on the well casing. The elevation of each RP for each well must be surveyed by a licensed surveyor to some pre-determined standard, such as the North American Vertical Datum of 1988 (NAVD 88). For wells that have been tightly capped and may be under pressure, groundwater elevations should be allowed to equilibrate to atmospheric conditions for at least 10 min before measuring. The sampler should use a measuring device capable of measuring depth to groundwater to an accuracy of 0.1 ft and should repeat measurements at each location to ensure accuracy.

When sampling for groundwater quality, sampling protocols should ensure that data are taken from the correct location and are representative of conditions that form appropriate basin management. Data should be recorded in a field logbook or other permanent documentation. Low-flow or passive sampling equipment may be used to minimize the volume of water to be extracted before sample collection (and, therefore, requiring disposal). Samples should be collected according to appropriate national and/or state standards.

When monitoring groundwater for saltwater intrusion, samples should be collected at least semi-annually (more frequently in areas of high groundwater extraction). Samples are typically analyzed for chlorides, though other parameters, such as TDS, may be used. Water levels should be monitored frequently enough to characterize changes in head near the leading edge of degraded water quality.

Monitoring may occur for other parameters at the same time as groundwater data are collected. For sample, streamflow data can and should be collected where questions regarding surface water–groundwater interactions are being asked. Recommended guidance for collecting, analyzing, and reporting streamflow data includes the USGS TM 3-A8.

Land subsidence data may also be collected in areas where declining groundwater elevations may result in inelastic subsidence. Unfortunately, no standard procedures exist for collecting and analyzing subsidence data; however, options include level surveying, continuous global positioning system (CGPS) data, static global positioning system (GPS) surveys or real-time kinematic (RTK) surveys, interferometric synthetic aperture radar (InSAR) data, and extensometer networks.

Maintenance of Monitoring Network

Monitoring is a long-term fluid proposition; as such, monitoring goals and networks should be re-evaluated regularly to ensure that the appropriate data are being collected. Data gaps should be reconsidered/identified and programs for filling those gaps updated. As part of the monitoring network updates, new or updated monitoring or sampling techniques and methods should be considered, and sampling and analysis plans should be updated to reflect current technology and industry standards. Equipment used in the program should be inspected and, if appropriate, recalibrated to ensure the accuracy of data.

DATA REVIEW AND TREND ANALYSIS

Data collected as part of monitoring programs should undergo review for quality control and accuracy and should be maintained in some form of a data management system (DMS). The Advisory Board for Water Information has developed a National Groundwater Monitoring Network and associated data portal and a water quality portal, both of which could provide important data for utilities looking to develop groundwater models.

Data Review

Following collection, data should be evaluated for how accurate, precise, complete, comparable, sensitive, and representative the water quality measurements are relative to the monitoring program's data quality objectives (DQOs). DQOs are established as part of the monitoring program by considering the specifications of the instruments, methods, and laboratories to be employed and by considering the utility of the data.

Accuracy. Accuracy describes how close the measurement is to its true value. Accuracy is the measurement of a sample of known concentration and comparing the known value against the measured value. The accuracy of chemical measurements should be checked by calibrating all meters and performing standard comparisons on instruments that cannot be calibrated, such as thermometers and temperature data loggers. The concentration of any standard will be within the mid-range of the equipment. A calibration record should be used to record equipment calibration and accuracy checks.

Comparability. Comparability is the degree to which data can be compared directly to similar studies. Standard operating procedures (SOPs) developed for the specific monitoring program or for a larger statewide or national monitoring program can be used.

Completeness. Completeness is the fraction of planned data that must be collected to fulfill the statistical criteria of the project. There are no statistical criteria that require a certain percentage of data be taken. However, it is expected that ≥80% of all measurements should be taken when anticipated.

For each parameter, breakdown of equipment, frequency of weather events, etc., may hamper completeness. At the end of the season, the number of samples collected will be compared to the planned number and the completeness will be presented as a percentage for each parameter. Reasons for not meeting the completeness objective will be recorded.

$$\%\ \text{Completeness (samples)} = \frac{\text{Samples collected}}{\text{Planned samples collected}} \times 100 \qquad \text{(Eq 5-1)}$$

Precision. Precision is the measure of how close repeated measurements are to each other. It describes how well repeated measurements agree. The precision objectives apply to duplicate and split samples taken as part of in-field quality control (QC) checks. Each piece of equipment or method of data collection will be repeated at a minimum of one station during each field-monitoring event. These repeated measurements may be performed at more than one site. If a sample result seems unusual or the procedures were questionable, take the measurement a second time.

Representativeness. Representativeness is the degree to which data truly characterize the actual environmental conditions. Sampling methods are designed to be as representative as possible.

Constraints to achieving representativeness can include:

- samples taken from a well with changing local land use or groundwater use in the monitoring well vicinity (e.g., adjacent intermittent pumping well);
- samples taken in a stream reach that does not describe the area of interest (e.g., an ambient water sample taken downstream of a point source);
- samples taken in an unusual habitat type (e.g., a stagnant backwater instead of a flowing portion of the creek);
- samples not analyzed or processed appropriately, causing conditions in the sample to change (e.g., exceeding holding times for laboratory processing); and
- access to sites because flooding conditions, closed roads, and private lands may limit representativeness for some sites.

Representativeness will be ensured by adhering to the established sampling and analysis methods.

Data Validation and Usability Elements

Data review, validation, and verification. Data sheets should be reviewed after each event by technical advisors to determine if the data meet the monitoring program quality objectives and to identify outliers, spurious results, or omissions. Compliance with the data quality objectives should also be evaluated. Corrective actions should be suggested for implementation in subsequent monitoring trainings and data collection events. Problems with the data quality and corrective action should be reported in final monitoring reports.

Validation and verification methods. As part of the standard field protocols, any sample readings out of the expected range should be reported to the designated monitoring lead. A second sample or reading should be taken as soon as possible to verify the condition. If data are invalid, then data should be noted and flagged on the data sheet. Further actions should be taken to trace the sources of error and to correct those problems.

Reconciliation with DQOs. If data quality indicators are not found to meet program specifications, data should not be entered into the DMS and should not be used in summarized annual reports. The cause of failure should be evaluated. If the cause is found to be equipment failure, calibration and maintenance procedures should be reassessed and improved. If the problem is found to be sampler error, the person should be retrained. If accuracy and precision goals are frequently not being met, a QC verification session should be scheduled for all persons collecting data.

If failure to meet program specifications is found to be unrelated to equipment, methods, or monitoring error, the specifications should be revised.

DOCUMENTS AND RECORDS

All field measurements and observations should be recorded on data sheets and/or field logbooks at the time of completion while in the field. Field personnel should also record and maintain instrument information in the instrument maintenance logs and/or calibration records.

Data sheets should be reviewed for outliers and omissions before leaving the sample site at the completion of each data collection event. Data sheets should be signed after review by a field monitoring leader. Following completion of the fieldwork, all data sheets should be sent to a data manager for cataloging and entry into the monitoring program DMS. Original copies of field records should be stored in an "original binder" and copies put into a "working binder." Copies of all information in the field logbooks should be made and inserted into the working copy binder. Entry of all data should be made into a computer database within six months of data collection. Computer backup copies should be made on an annual basis. All data entry and other tasks involving data sheets should utilize the working binder. The original binder should be used as a reference only. Additionally, it is recommended that field sheets are archived for a minimum of three years from the time they were collected.

Instrument maintenance and calibration records should be kept by field personnel for each instrument in use. These records should include data for the temperature data logger units, field thermometers, and water quality kits and meters. The instrument records detail the dates of equipment inspection and calibration, as well as the dates reagents are replaced. Instrument records should be turned in with data sheets and photocopies placed in the working binder. A field site log pertaining to sampling locations, including maps, directions to sample sites in the field, photographs, and site characteristics, should also be maintained as part of the sampling records and updated, as needed or as site location changes.

Hydrogeologic Databases

In developing a monitoring program, existing hydrogeologic databases and statewide or national monitoring programs should be consulted. Many publicly available databases are available online and contain a wide range of hydrologic and hydrogeologic information. The following is a list of typically available database sites with web links:

Geologic data

- USGS National Geologic Map Database (http://ngmdb.usgs.gov/ngmdb/ngmdb_home.html)
- USGS Publications Warehouse (https://pubs.er.usgs.gov/)
- state and regional databases
- oil and gas data

Surface water data

- USGS National Water Information System (http://waterdata.usgs.gov/nwis)
- state and regional databases

Climatic/weather data

- NOAA, National Weather Service (www.weather.gov/)
- National Weather Service Climate Prediction Center (www.cpc.ncep.noaa.gov/products/predictions/long_range/seasonal.php?lead=5)
- rainfall/weather data
- state and regional databases (Western Regional Climate Center, www.wrcc.dri.edu/)

Groundwater data

- USGS Groundwater Information (http://water.usgs.gov/ogw/data.html)
- USEPA STOrage and RETrieval (STORET)
- USEPA Water Quality eXchange (WQX)
- UEPA Safe Drinking Water Information System (SDWIS)
- state and regional databases
- other utilities
- submittals for regulatory purposes/reporting

REFERENCES

ANSI/AWWA. 2015. A100, Water Wells. Denver, Colo.: AWWA.

AWWA. 2014. (4th ed.). M21, Groundwater. Denver, Colo.: AWWA.

Bloetscher, F.; Muniz, A.; and Largey, J. 2008. *Siting, Drilling and Constructing Wells.* Denver, Colo.: AWWA.

Chapter 6

Groundwater Models

Computer-generated models of groundwater behavior, termed *groundwater models* in this chapter, have been used for decades to answer questions pertaining to the movement of groundwater flow and contaminant transport. Groundwater models are most typically used when the level of impacts resulting from implementation of one or more groundwater-related projects is not well understood and could result in irreparable harm to the groundwater basins and connected water resources. To construct the model, the monitoring, lithology, water quality, and other data, as have been described in previous chapters, must be gathered and entered into the model. Putting all the data together to create a computer model of the aquifer can help the utility with its long-term plans for management of the resource. As a result of recent advances in data collection methods, computing power, university programs focused on groundwater modeling training, and visualization software have led to more regular use and acceptance of groundwater models by water supply managers and their consultants. This change has also been driven by increases in the cost and complexity associated with extracting clean water on a long-term sustainable basis from wells drawing from aquifers that could potentially become compromised by anthropogenic or naturally occurring contamination or by long-term decreases in aquifer levels. In addition, as water supply demands and aquifer stresses increase with time, the potential for other undesirable impacts such as saltwater intrusion, upconing of poorer quality water, aquifer depletion, and land subsidence increases. Furthermore, increased regulatory requirements and the potential for adverse environmental impacts, such as stream depletion because of groundwater extraction, have provided further context for the use of groundwater models to help engage all stakeholders involved in a groundwater management project and to evaluate the impact of groundwater and related projects on the interaction of complex hydrologic systems. Simply put, a groundwater model—when appropriately designed to address the project objectives—can save a utility significant time and money and can be a powerful tool in the sustainable long-term management of a groundwater basin.

This modeling can take the form of analytical solutions for simple cases or numerical computer codes for more complex cases where a high degree of accuracy is needed. All groundwater utilities should consider development of working computer models of their aquifer formation, calibrated regularly to collected data. MODFLOW, developed by the US Geological Survey (USGS) (McDonald and Harbaugh, 1988, 1984), is the basis for most commercial groundwater models today, especially those medium and large utilities that will exert greater impact on water demands. MODFLOW may be sufficient for relatively simple local applications, but more extensive modeling and significantly more data may be needed if surface interfaces and competing users are located in a given basin. The basin approach, even with groundwater, appears to be more appropriate than simple local modeling. There are numerous resources available that describe, in varying levels of detail, the processes, equations behind, and numerical methods of groundwater modeling, including those listed in the reference pages. A fully developed derivation of these equations can be found in Bloetscher et al. (2005). Therefore, this chapter does not seek to replicate the level of detail in those references but instead aims to be a succinct reference document for those considering the use of, or evaluating the results from, a groundwater model. The hope is that the content included in this chapter, used with the case studies in Chapter 7, can be used to answer the following questions:

- Will the use of a groundwater model add value to the project?
- Is the use of the proposed groundwater source sustainable?
- What level of model complexity/effort is needed to satisfy the project objectives?
- Which type of model is needed to meet the project objectives?
- Are the data currently available sufficient to support the level of complexity required to answer the questions posed within a reasonable amount of uncertainty?
- How should a groundwater model be reviewed to ensure that the results are understood and that the recommendations made are based on best principles and a thorough understanding of the model's limitations?
- What factors should be considered when deciding how long-term stewardship of a groundwater model should be undertaken?
- Is the use of the proposed groundwater source reliable?

These questions are among the most relevant to utilities because they get to the heart of understanding a groundwater model's potential value and the information needed to effectively communicate with those doing the modeling. Unlike engineering designs where precise measurements are turned into something tangible, a groundwater model attempts to "approximate reality" while being transparent regarding its uncertainties and limitations over a potentially open-ended time frame. As such, a groundwater model can "live" for decades as an efficient tool for assessing current and future questions, but the acknowledgment of the uncertainties is a critical component of long-term management and decision-making. At the same time, it must be understood that models are replications of typical or average conditions estimated for the aquifer parameters. Hence, there is also a need to understand the limitations of these models.

WHEN TO USE A GROUNDWATER MODEL AND WHY

The use of a groundwater model should be considered when the absence of one would impair a utility's ability to allocate available resources (time and money) in a manner beneficial to the stakeholders of a groundwater management project. In this context, a model could be simple analyses that are more efficiently completed using an analytical model with hand calculations or a detailed numerical computer model to address complex questions that are too multifaceted to attempt without the assistance of a computer. Conversely, if data limitations, site complexity, and/or the presence of a limiting process that cannot be described numerically prevent the model from producing precise enough results to satisfy the objectives of the project (because of high levels of uncertainty), other means of analysis should be considered.

TYPES OF GROUNDWATER MODELS

Groundwater modeling is an umbrella term that incorporates many types of software and practices. At the most fundamental level, a groundwater model is any representation of an aquifer system that is used to better understand how the system functions. This can include:

- nonsimulation types of models, such as 3-D visualization models, where data and structural interpretations are input to a 3-D environment that can be examined through a viewer. These models have no predictive capabilities but can be a powerful tool in identifying data gaps and areas where more system understanding is needed. A 3-D visualization model can also be a precursor to the development of a numerical computer model;
- decision support models that quantify the transfer of water based on operational and system constraints are used to optimize system usage or to guide utilities toward a preferred set of conditions. These models are not physically based (e.g., they do not solve the equation of groundwater flow) and largely seek to solve mass balance equations under a range of imposed conditions (e.g., drought);
- analytical models, where mathematical equations are solved to describe phenomena such as drawdown in a well, are useful tools to help develop an understanding of system characteristics. While they can be used in a limited capacity for predictive means, they offer limited ability to incorporate multiple stresses (pumping, river discharge), aquifer heterogeneity (varying hydraulic properties), or other spatial (in planar or vertical) variability; and
- numerical computer models of varying complexity, which are often used to predict aquifer responses to stresses and develop plans for future operations and/or development. Numerical computer models use the methods described in Bloetscher et al. (2005) to discretize an aquifer domain into small pieces. Using these smaller increments, the physically based equations governing the process at hand are solved.

It is common for a combination of these types of models to be used. Analytical and decision support models can be used to represent a specific process or management plan, with the results converted to input to a numerical model.

Numerical models are broadly divided into two types, based on the method used to discretize the solution domain and the structure of the grid; these are referred to as finite difference and finite element models. The details pertaining to each method are described at length in Anderson et al. (2015) and will not be repeated here. The major differences are summarized as follows.

- When viewed in plan view, standard finite difference grid cells, such as the ones employed in MODFLOW-2005, are rectangular, as shown in Figure 6-1. Each grid cell is associated with a row, column, and layer, and the addition of grid cells is facilitated by adding rows, columns, or layers to the domain. The equations are solved for in either the center of the cell (block-centered) or at the intersections (edges) of the cells (mesh-centered).
- Finite element grid elements (equivalent to cells in finite difference) are triangles or quadrilaterals in plan view. Finite elements allow more flexibility in grid design as elements can be added anywhere in the model without corresponding to a row or column. Equations are solved at nodes, which reside at the intersections of the elements.

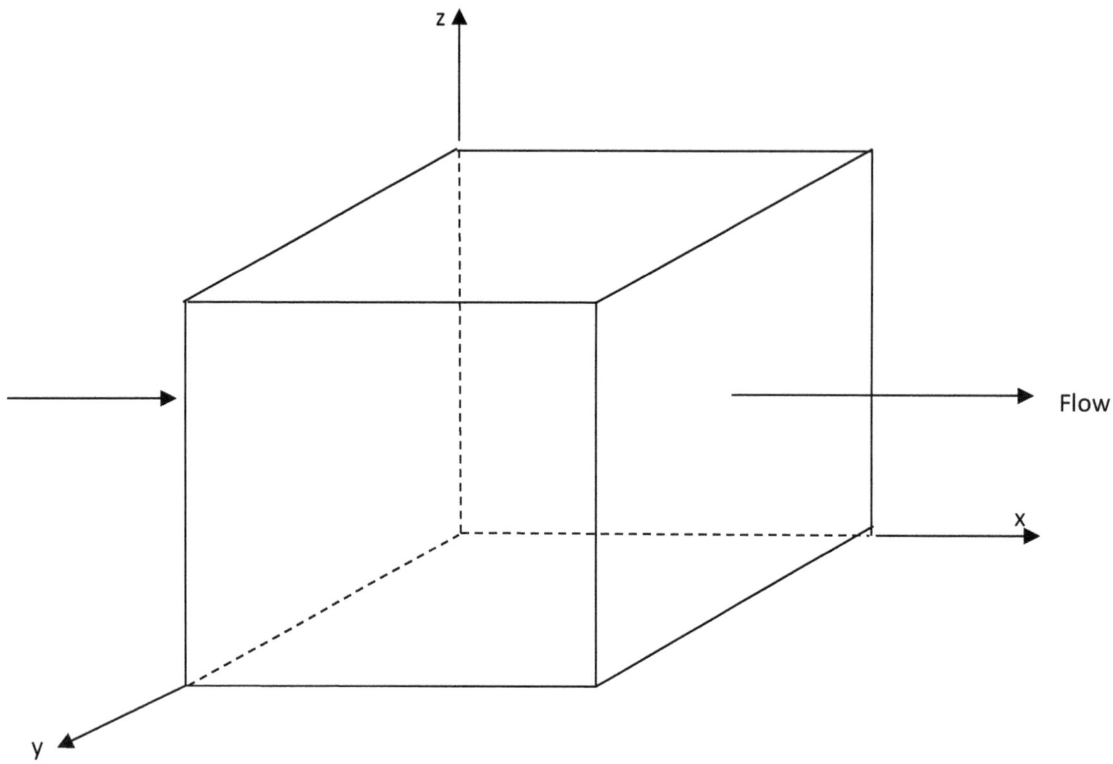

Source: Created by author

Figure 6-1 Block-centered flow regime

Because of advances in computing power, models can now incorporate a considerable number of nodes and elements/cells. Both solution methods produce equivalent results, so the utility staff should consider the relative need for grid flexibility and the experience of the modeling team with either type of code when selecting a numerical model type.

Types of Modeling Software

Groundwater modeling software can be grouped as follows:

- model code – the software that solves the equations for groundwater flow or any other process;
- graphical user interface (GUI) – visualization software packages that allow users to see the model inputs, structure, and results on the computer screen;
- Pre- and post-processing software – software, such as databases, geographic information systems, spreadsheets, scripts, etc., that is used to collate, organize, and convert raw data into model input or further analyze model output; and
- optimization/calibration software – software that supports automatic model calibration, sensitivity analysis, and uncertainty analysis.

Regularly used software products from each of these groups can be found that are either public domain and freely available for use and modification or commercially available and only available in executable form that cannot be modified by the end user. Some popular model codes, such as MODFLOW-2005, are public domain but are often used along with GUIs that are available only commercially.

Selection of an appropriate model depends on the goals of the modeling exercise. If the model is being developed only for water supply, a flow model may be sufficient. If saltwater is an issue, a standard flow model will not capture the density differential issue with saltwater, so a more advanced model, such as USGS's SEAWAT, will be needed. Contaminant transport models add to the complexity of the model. They rely on the flow models being developed correctly to make projections as to the movement of contaminants toward water supply wells. Particle tracking can be used to complement contaminant transport models (MODPATH by USGS). Risk models add a new dimension; they require that contaminant transport models appropriately represent aquifer conditions so that the contaminant concentrations can be translated to risk contours (all of these will be used in Chapter 7). In addition, each requires additional data to represent the aquifer conditions properly.

GROUNDWATER MODEL COMPLEXITY

Groundwater modeling tools can vary greatly in level of complexity and the amount of resources in terms of time and money needed to develop and implement. The model(s) chosen should represent the simplest approach that will meet the project objectives. The following questions should be asked when assessing the level of complexity needed to satisfy the project objectives:

- How many spatial dimensions are needed to represent the system?
- Can the system be represented as a porous media?
- Can the system be represented as "steady-state," with model inputs and outputs constant over time?
- How precise does the model output need to be to address the questions posed in the project objectives?
- Which processes need to be represented in the analyses?
- Has a groundwater model already been developed for the project area?

Model Dimensions

Groundwater models can be one-, two-, or three-dimensional space. One-dimensional models are typically used to understand one specific process and often employ analytical solutions to solve mathematical equations governing the process in question. Some examples include the following:

- The calculation of drawdown at a pumping well (or nearby monitoring wells) based on aquifer properties (hydraulic conductivity, storage/specific yield) and well properties (pumping rates, well radius, full/partial penetration) is employed regularly to analyze aquifer performance and slug tests.
- Solute transport along a single flowpath based on aquifer properties (groundwater velocity, effective porosity, hydraulic conductivity, dispersion) and solute properties (sorption, decay) would be used to quickly predict the longevity (time) and extent (space) of a plume.
- Movement of water applied at the ground surface through the vadose zone based on vadose zone properties (water content, drainage parameters, thickness), and application rates is used to approximate groundwater table mounding beneath a recharge basin or other source of water.

- Diffusion of a solute into/out of bedrock based on fracture rock properties (frequency and size of fractures) and solute properties is used to approximate the longevity of solute mass trapped in inaccessible rock pores.

Two-dimensional models are often employed when the system in question does not exhibit significant variations in a third spatial direction. Examples include the following:

- Planar (*X* and *Y* direction) representation of groundwater flow or solute transport in an aquifer that has very small vertical head gradients may be employed to estimate pumping well zones of contribution/capture or potential discharges to other receptors (drains, water bodies, sumps, etc.).
- Cross-sectional (*X* and *Z* direction) representation of groundwater flow or solute transport may be used to understand processes such as groundwater table mounding because of recharge application, saltwater intrusion in a coastal aquifer, or multiphase flow (such as Dense nonaqueous phase liquids (DNAPLs) movement) governed by both the groundwater flow field and aquifer geometry.

Three-dimensional models are most often used to represent groundwater flow regimes that have a higher level of complexity, often including multiple aquifer/aquitard layers, significant vertical gradients, multiple discharge locations, mechanisms (pumping well, stream discharge, etc.), and depths, and spatially varying hydraulic properties (heterogeneity). Examples include:

- optimization of a well field for water supply to maximize yield and minimize adverse impacts to sensitive receptors,
- delineation of wellhead protection areas,
- solute transport analyses to delineate a groundwater contaminant plume in three dimensions in support of a remedial design or monitoring natural attenuation recommendations,
- screening of enhanced stormwater recharge (through low-impact development [LID] or similar approaches) basins/locations to predict water table mounding on a regional (citywide) or local (area surrounding a specific LID installation) scale,
- estimating sustainable yield of a coastal aquifer with respect to saltwater intrusion, and
- determining the feasibility of aquifer storage and recovery (ASR) projects.

Porous Medium Representation

Most unconsolidated aquifer systems transmit water according to the mathematical laws and equations governing fluid flow through a saturated porous medium. These equations conceptualize a system of directly interconnected pores saturated with one fluid (typically water) that moves from areas of high pressure/head to low pressure/head at velocities that depend on the properties of the aquifer (hydraulic conductivity and effective porosity). All movement is based on the application of Darcy's Law, which states that the volume of water that can be withdrawn from an aquifer is the hydraulic conductivity, multiplied by the aquifer area and multiplied by the aquifer gradient. Henry Darcy developed this relationship while conducting studies on how media filters work. He noted that the quantity of flow was proportional to the hydraulic conductivity, area, and slope. See Appendix A for more discussion of Darcy's Law.

Systems that do not have directly connected adjacent pores, such as fractured bedrock, can be represented as a porous medium under specific conditions. Using fractured bedrock as the most common example, the following should be considered:

- The (fractured rock) system may be able to be represented as an equivalent porous medium (EPM) at large (regional) scales where bulk properties can be estimated and groundwater flow gradients can be discerned.
- As the scale of the study area decreases, flow is more likely to be governed by the specificity of the local fracture network. In these cases, an EPM model is not appropriate, and a more complex discrete fracture network (DFN) model would have to be used.
- The data required to support and minimize the uncertainty of a DFN model are often substantial. Utilities should exercise caution when embarking on a groundwater modeling project in a fractured rock system.

Temporal Representation

Groundwater models are typically represented as either "steady-state" or "transient." In a steady-state model, the inputs to the simulation are assumed to be constant with time. When simulated, this produces one result indicative of a solution that represents the inputs to the model. For example, the use of average groundwater pumping, average groundwater recharge, and average boundary heads (the water surface of a lake, for example) would produce groundwater head equipotential contours representative of "average" conditions. For many groundwater applications, determination of average conditions is sufficient, and this lower level of complexity is warranted. However, transient analyses may be required if the following apply:

- There is significant variation between average conditions and what typically occurs during a representative time period.
- The answers sought require an understanding of the rate of change of a process (e.g., the change in heads over time corresponding to pumping).
- The inputs used are unbalanced to a degree that a realistic equilibrium is not attainable. For example, an input of zero recharge and average pumping to a transient model could show groundwater heads dropping with time and, depending on the aquifer properties and boundary conditions, could produce a result that is both realistic and sustainable over a discrete time period. The same inputs to a steady-state model would likely dewater the aquifer, rendering the results useless.

To conduct transient modeling, a "time step" or "stress period" (these terms have slightly different meanings, depending on the model code used, but are assumed to be interchangeable in this manual) is assigned to the model along with start and end times/dates. For example, a model could simulate the period of 1/1/2000 through 12/31/2009 (10 years) using monthly time steps. If that were the case, 120 flow field results would be produced corresponding to the inputs provided for each time step. Most model codes can use a variable time step (e.g., 10-day time steps from 1/1/2000 through 12/31/2000, then monthly time steps through 12/31/2009) and can average data within a time step (e.g., producing a single, average pumping rate for December 2009 given daily pumping rates as input).

Typical time-varying inputs to a transient model include pumping, recharge, boundary heads, and boundary fluxes. Most groundwater models assume that properties associated with the solid system such as hydraulic conductivity and ground surface elevation are unchanged with time. More sophisticated models incorporate land subsidence or changes related to excavation and/or backfilling over time.

Structural Model Detail

A numerical computer groundwater model typically involves the creation of a fixed grid or mesh (term used interchangeably in this manual) within a fixed domain. While variations between different model codes exist, the codes all have the following general functionality:

- The planar area encompassed by the model domain is subdivided or "discretized" into user-defined sections referred to as cells or elements. These elements can be any type of shape, depending on the codes used, but are typically rectangles (for finite difference models) or triangles (for finite element models).
- At the intersecting points or center points of each cell (also called nodes), the equations of groundwater flow (and any other process simulated) is solved for each specified time step. This nodal output is what is used to generate head equipotential, drawdown, and contaminant concentration contours that are usually presented by groundwater modelers.
- Point-specific inputs, such as pumping from a well, are applied directly to the corresponding nodes, while area-specific inputs, such as hydraulic properties and recharge, are applied to model cells.
- Along the boundary of the active domain, conditions are specified that govern flow at these extremities. Typical conditions include zero flux (also known as "no flow") corresponding to a groundwater divide; specified head (also known as "constant head," though this can be a misnomer as the specified head does not necessarily have to be constant in time); specified flux; and a general head boundary (GHB), which imposes a mixed condition where components of specified heads and fluxes are applied. Regardless of which condition is specified, preference should be made to select boundaries that correspond to "natural" features such as mountains (zero flux, presumed groundwater divide) or large water bodies (specified head set to water body water level). In the absence of such natural features, assumptions must be made.
- Vertically (if it is a 3-D model), the model is divided up into layers, which typically have the same planar grid imposed on them and are designed to roughly correspond to the stratigraphy of the system. The tops and bottoms of each layer are referred to as levels or slices, depending on the nomenclature of the code, and nodal calculations are made at either the center of the layer or at the level or slice.

The amount of detail required for a model is related to the detail of the available data, the natural features present, and the objectives of the project. As computations are made at each node, it is typically desirable to have more nodes (and, therefore, smaller grid spacing or more discretization) in the areas of the model domain most important to the system and/or where steep gradients are present. These typically include pumping wells and water bodies where groundwater discharges. While there are no agreed-upon guidelines on grid spacing or model domain selection, the information desired from the model or permitting requirements for the model may delineate what the parameters need to be. Likewise, the model should be evaluated to ensure that it is capable of providing the needed guidelines (the ability to do so varies).

More detail is likely to correspond to higher levels of effort and costs because of the computational requirements and corresponding longer run times associated with an increase in discretization. Care should be taken to ensure that grid detail is appropriately targeted to the areas of interest. As the number of nodes in a model is calculated by multiplying the nodes included in the planar grid by the number of levels or slices, the addition of model layers should be done on an as-needed basis.

The amount of detail included in the assignment of hydraulic properties should be dictated by the available data, the conceptual model, and the model calibration process.

When possible, the simplest approach should be prioritized. The same applies to the assignment of rainfall recharge zones.

Model Processes

Present-day model codes (as a whole) can represent nearly every process imaginable. Care must be taken to match the processes selected to the objectives of the model. The most typical processes used are:

- groundwater flow – simulates piezometric heads and velocities at computational nodes;
- particle tracking – simulates the forward- or backward-in-time movement of a water particle released within the model domain, based on the simulated groundwater flow field. This process cannot be used to represent dispersion or decay, though it can include linear sorption (also known as retardation) and produces pathlines and associated flow direction arrows only;
- dispersive solute transport – simulates the movement of mass input to the model incorporating both advective (flow direction and velocity, which can incorporate retardation) and dispersive (dispersion and decay) mechanisms and uses the simulated groundwater flow field. Typical outputs include simulated concentrations at nodes associated with each time step;
- sharp interface – simulates the interaction of fresh- and saltwater by approximating the transition between the two zones as a sharp interface (Figure 6-2). This approach is valid and favored for regional aquifer analysis where the utility is interested in determining sustainable pumping rates with respect to saltwater intrusion;

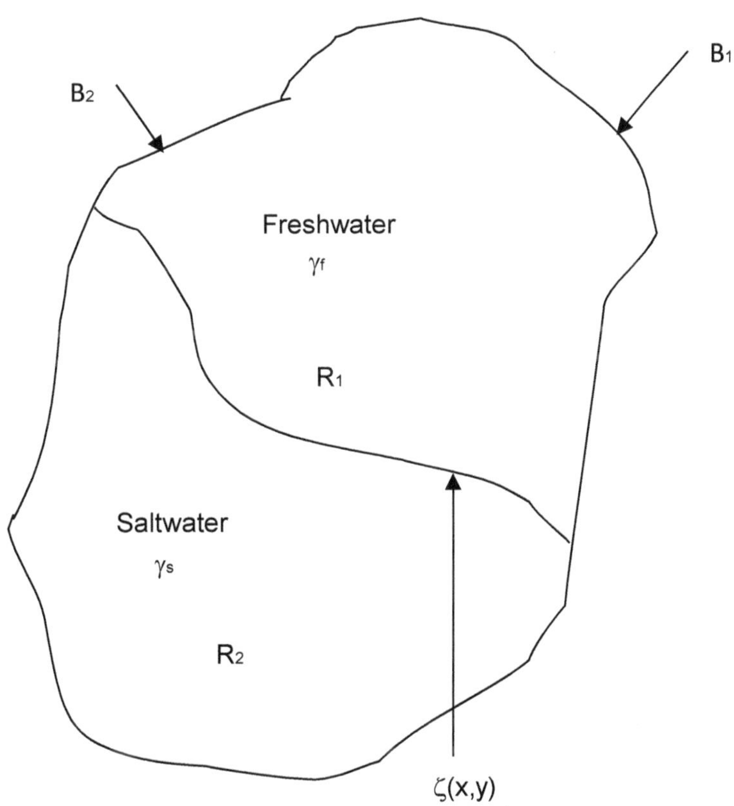

Source: Created by author

Figure 6-2 Saltwater–freshwater interface

- variable density solute transport – incorporates differences in densities that would be encountered between fresh- and saltwater (in a coastal aquifer or subsurface brine system) into dispersive solute transport simulations. This approach is more complex than a sharp interface analysis and is used when simulating chloride concentrations, which is necessary to meet the project objectives;
- reactive solute transport – simulates the dispersive solute transport and incorporates interactions and transformations of chemical species, such as the transformation of perchloroethylene (PCE) to daughter products;
- heat transport – simulates the movement and dissipation of heat in an aquifer system, which is predominantly used in underground geothermal energy studies and designs;
- multiphase flow – incorporates more "phases" in the simulation than simply water (liquid phase) and soil (solid phase). This can include simulation of flow through the vadose zone that incorporates air (or other vapors) as a gaseous phase or the simulation of non-aqueous phase liquids (NAPLs) such as the free product of chlorinated solvents (e.g., PCE) that are prevalent in contaminant spills; and
- geochemical interaction – incorporates the geochemical interactions between the soil and the solutes to approximate transformations resulting from in-situ oxidation and/or reduction.

Additional Complexities

The utility should be aware of additional complexities (beyond the processes represented) that are inherent in the development and use of groundwater models. A subset of the most common of these includes the following:

- representation of the water table: Different model codes represent the water table and its changing position over the course of a simulation differently. Specifically, when the water table drops below the bottom of the topmost computational layer in a 3-D numerical model, computations must be made to accommodate that "dry" layer. Each code handles this situation differently, and code limitations can bias the design of a model to avoid numerical instabilities. The utility staff should ask the modeling team how it plans to handle water table variations in the model during the model conceptualization stage to ensure that the right code is chosen, and the model is designed and constructed prudently and without code-related bias.
- Wells that pump across more than one stratigraphic layer: Pumping wells with screens that span more than one stratigraphic layer add complexity to a groundwater model in two ways. First, the division of flux between stratigraphic layers is often not known and is usually approximated. Second, if significant head differences exist between the stratigraphic layers, the well may act as a conduit of flow (in the direction of the vertical gradients) when it is not pumping. The utility staff should ask a modeling team how it plans to represent such wells in the numerical model during the model conceptualization stage.
- Well losses: Groundwater models are commonly used to simulate drawdown associated with pumping wells. This drawdown represents the change in head within the aquifer associated with the stresses applied by the well and is a function of aquifer geometry, aquifer properties, well screen length and placement, and pumping rates. The equations employed by groundwater models to perform these analyses do not include head losses (and associated drawdowns) resulting from well losses. As a result, two things should be considered during the conceptualization

of a groundwater model. First, observed drawdowns at a pumping well should be higher (because of well losses) than simulated drawdowns at the well. Second, model calibration efforts related to changes in head because of pumping should focus on changes in heads at nearby monitoring wells and not within the wells being pumped.

- Surface water–groundwater interaction: This includes any situation where water above the land surface interacts with groundwater below the land surface. This can be as simple as understanding where groundwater is expected to discharge to surface water bodies to as complicated as representing storm-specific watershed runoff and associated streamflows and interactions with groundwater aquifers. The level of complexity employed should be tied directly to the objectives of the project and should be discussed during the conceptualization phase. A thorough description of this topic is included in the Australian Groundwater Modeling Guidelines (Barnett et al. 2012).
- Boundary fluxes: As noted previously, natural boundaries should be used in a groundwater model whenever possible. When this is not possible, care must be taken to estimate the fluxes that cross any "artificial" model boundary and to ensure that the artificial boundary is not too close to pumping wells or other sources or sinks as to influence the outcome of the simulations. One such example is the representation of a valley aquifer system, with the boundary placed at the base of a hill or mountain range. A portion of the flow contributing to the aquifer system will be from recharge occurring within the hills or mountains, outside of the model domain. A thorough water balance of the area contributing to the model boundary must be completed before model development to ensure that this flux is not over- or under represented.
- Representation of faults: Faults can represent a discontinuity in the subsurface flow regime that can manifest itself in many ways. Some faults create a barrier to flow, while others are relatively benign. Faults should be characterized in the context of groundwater flow before model development.
- Linking different models together: Linking simulation codes together is sometimes necessary and often complicated. Care should be taken to understand the limitations of the software with respect to the proposed linkages before software is selected and purchased.

THE MODELING PROCESS

The process of developing and utilizing a groundwater model is designed to methodically build an understanding of how the aquifer system works by iteratively updating the conceptual and simulation models based on their relative fit to the project data. When done properly, the end product is threefold:

- a refined conceptual model of how the aquifer system works, the inflows and outflows to the system, and the relative impacts of stresses on the system;
- a simulation model that can be used to test alternatives and help with the development of future operational plans; and
- the development of expertise on the model, the aquifer system, and the data set by the model team members who can be used strategically to collaborate with, train, or present alongside the utility staff.

Given this preferred triad of outcomes, the following steps are usually taken during the modeling process:

- Step 1: Identify objectives
- Step 2: Assess available data
- Step 3: Develop a conceptual model
- Step 4: Build the numerical model
- Step 5: Calibrate the model
- Step 6: Test the reliability of the model through a sensitivity analysis
- Step 7: Utilize the model to meet the objectives
- Step 8: Documentation and long-term stewardship

While presented in linear fashion, steps 3–6 are often iterative, and step 8 (documentation) should be ongoing throughout the entire modeling process. Clear communication of the model development and output through meetings, workshops, and reporting allows the owner, outside parties, and the modeling team to collectively assess whether the model is appropriately meeting the project objectives. Each step is described in detail in the following sections.

Step 1: Identify objectives

The objectives of the modeling work should be agreed upon before progressing to any other step in the modeling process and should be reviewed and discussed regularly as the project moves forward. The objectives should provide a framework for how the model can specifically contribute to answering the questions being posed by the utility or other stakeholders. Once the objectives are established, the type and complexity (and subsequent confidence level / uncertainty) of the model can be proposed and discussed. At this stage in the process, the goal should be to clearly communicate and agree upon the expectations of the modeling team, the utility, and stakeholders, which may include competing users, regulatory staff, and numerous other interested parties.

Step 2: Assess available data

The amount, spatial and temporal coverage, and quality of the available data to be used to meet the project objectives must be assessed before commencing with modeling activities. Some examples are as follows:

- Objective: Predict piezometric head drawdowns from a range of potential operating scenarios.
- Objective–specific data need: Time-varying piezometric head data, showing the historical or test-based response within the aquifer to changes in pumping rates, would be needed to verify that the model is reliable.
- Objective: Predict the impacts to stream baseflow from groundwater pumping.
- Objective–specific data need: Streamflow data would be needed to verify that the model represents baseflow accurately.
- Objective: Predict whether or not a contaminant plume will impact a supply well and, if so, at what concentrations.
- Objective–specific data need: Information on the source, historical pathway, and present-day concentrations of the plume, plus data on all potential sinks (supply wells, industrial wells, drains) in the area.

Some types of data are nearly always needed to build and calibrate a groundwater model, regardless of the objectives. These include:

- data indicative of the hydraulic properties of the materials, such as aquifer performance tests (preferred), single well pumping tests, or slug tests;
- a synoptic round or rounds (preferred) of piezometric head data taken from observation wells with known measuring point elevations and screen interval elevations;
- land use and land cover data;
- rainfall data;
- topographic data;
- pumping well locations, screened depths, and rates;
- lithologic (or similar) well logs from which stratigraphy can be inferred; and
- the location of any other feature(s) that may add or remove water from the system.

Early collaboration and assessment to determine if the available data are sufficient to meet the project objectives is beneficial. Identifying and addressing data gaps before the commencement of model activities can save time and resources in the long run by ultimately reducing the uncertainty in the model predictions.

In some cases, this step in the process is the first time all of these relevant data are assessed together and compiled in a project database. Utilities should look for opportunities to add value to the project by requesting that the compiled data be formalized and delivered in a manner that can be used further going forward.

Addressing data gaps is a critical complement to this task. Knowing what data are needed and what sources might need to be reviewed or what test well data might be needed will help the utility with developing the model.

Step 3: Develop a conceptual model

A conceptual model is an understanding of how the system works with respect to the water balance, imposed stresses, sources and sinks, hydraulic properties, stratigraphy, boundaries, and groundwater flow gradients. The conceptual model initially serves as the roadmap for the design of the numerical model but should be updated iteratively during the model development and calibration phases of the project. It is good practice for a groundwater modeling project to have a deliverable or workshop once the conceptual model is formulated. This allows the utility and stakeholders to review and provide feedback before numerical model development. It is typical to include the following in the documentation of a conceptual model:

- a map with the boundaries (including sources and sinks) of the hydrologic system,
- a tabulation of the water balance,
- relevant surface water features,
- cross sections showing stratigraphy based on well/boring data,
- hand-drawn piezometric head contours with groundwater flow direction annotations based on synoptic rounds of data,
- maps depicting hydraulic property estimates based on available data, and
- time series plots illustrating the temporal variability of the system.

Once the conceptual model is completed, steps 1 and 2 should be revisited to confirm that the available data, as compiled in the conceptual model, are sufficient to meet the objectives of the project. If the data are not sufficient, the conceptual model should be used to identify data gaps and propose further data collection activities.

Step 4: Build the numerical model

In this step, the data compiled as part of steps 2 and 3 are converted to input files for the model. If a numerical model is used, this step requires the definition of a model domain, the creation of a numerical grid (or mesh, used here interchangeably), and the interpolation in space onto the grid of the following:

- ground surface elevations,
- stratigraphic contact elevations,
- hydraulic properties,
- rainfall recharge zones (often based on land use and land cover data), and
- return flow zones (based on whether an area is using septic systems or sewers).

Boundaries, sources, and sinks are then assigned to the model domain. Outer boundaries are typically either specified head, specified flux, or a mix of the two. Internal (i.e., not along the edges of the model domain) sources and sinks of water are typically pumping or injection wells, drains, rivers, and recharge.

Numerous methods are used to build a groundwater model. Fundamentally, this step is simply about turning data into model input, and the details of how it is done are not essential to this discussion. Once built, the model should run without errors and produce a stable and convergent solution.

Step 5: Calibrate the numerical model

Model calibration occurs when the modeler attempts to verify that the model replicates historical results. Model calibration entails both the adjustment of parameter values within their range of uncertainty and the investigation into and understanding of anomalous data to produce a model that can both represent historical data and be used with confidence to predict system responses to stresses applied in the future. Calibration is, by nature, an iterative process, often involving communication of unexpected system responses with field teams that can ground-truth modelers' intuition. What is gained through the calibration process is not only the model as a product to be used but a better understanding of how the system operates and what the system is sensitive to.

Calibration of a numerical model can be done heuristically (also known as trial and error) by manually adjusting values based on feedback from the previous simulation's results, or automatically by using software that automatically adjusts parameters based on feedback from the previous simulation's results, or a combination of the two. There are benefits and tradeoffs for all three approaches. It is generally good practice to have a heuristic component of the calibration process to ensure that the modeling team develops an understanding of the system, the data, and the sensitivities before ceding the processes to an automated computer algorithm. For example, a modeling team is considerably more likely to identify, understand, and determine how to move forward with data outliers when calibration is performed heuristically. An outlier could exist because of poor record keeping (perhaps the measuring point elevation is in the wrong vertical datum), system heterogeneity, or it could be indicative that the conceptual model needs revision. Going down the path of automated calibration without first developing this understanding can (and often does) lead to wasted effort and unrepresentative results.

The end result of model calibration is often the following:

- plan view maps showing "residuals" (simulated differences between simulated and measured values) at each well spatially. These plots should cover each stratigraphic unit individually and should provide an understanding of the spatial distribution of the "goodness of fit" or representativeness of the calibration. They can also highlight data outliers (for example, if three wells adjacent to each other have varying residuals, more attention is likely needed to understand this discrepancy). These residuals are also usually tabulated as well;

- a scatter plot chart comparing simulated and measured values to a "perfect fit" (a 45-degree line);
- statistics including, but not limited to, the following:
  - *mean error* – the average difference between measured heads and simulated heads;
  - *mean absolute error* – the mean of the absolute value of the differences in measured and simulated heads.
  - *root mean squared error (RMSE)* – the average of the squared differences in the measured and simulated heads; and
  - *scaled or normalized RMSE* – the RMSE normalized by the range of head values measured within the area of interest (which could be the entire site, the stratigraphic unit, or any other agreed-upon grouping of the data); and
- hydrographs or time series plots that compare the goodness of fit over time for a transient simulation. In a transient calibration, there will likely be several rounds of synoptic data for which residuals can be plotted as noted previously. However, the time-varying trends cannot be visualized on such a static plot. The hydrographs provide a way of quantitatively and qualitatively assessing the calibration. In some instances, it may be helpful to plot hydrographs on plan maps. These plots would help visualize the spatial distribution of the match to temporal variations in the system.

There is no uniform best practice or guidance in how to assess the goodness of fit of a model calibration nor should one be created and imposed on all projects. The goodness of fit will depend not only on the skill and judgment of the modelers but also on the types, spatial and temporal distribution and quality or level of uncertainty inherent in the data, the heterogeneity of the system, the level of complexity of the system being modeled, and the critical targets needed to meet the project objectives. In some instances, resources spent improving calibration statistics could be better spent, usually because of the law of diminishing returns.

However, no calibration should be deemed complete without a thorough documentation and understanding of the water balance. Calibration is inherently a nonunique process with nonunique outcomes. In the context of a groundwater model, this typically means that you can achieve a satisfactory goodness of fit with a steady-state model through the use of a near-infinite combination of flux into the model and hydraulic conductivity values assigned within the model cells. Simply put, imposing high flux and high hydraulic conductivity on a model domain can produce similar piezometric heads to that of imposing low flux and low hydraulic conductivity. For this reason, it is imperative that final calibrated model water balances be examined in the context of the conceptual model and deviations be examined and justified. This is of paramount importance if the objective of the project is to understand the relationship between potential future pumping rates and hydraulic capture.

To reduce the uncertainties associated with nonuniqueness, it is good practice to include a transient component in the model calibration. In the context of the preceding paragraph, while the piezometric heads across the majority of the model domain could be similar, drawdowns in the vicinity of extraction wells will be more sensitive to changes in hydraulic conductivity than boundary or distributed (such as recharge) fluxes imposed on the model domain. For example, if a model with high flux and high hydraulic conductivity matches drawdowns from an aquifer performance test well, the version of the model with low flux and low hydraulic conductivity will overpredict drawdowns, and vice versa. If possible based on available data and resources, a historical time period should be used as the basis for transient model simulations. If no such historic data are available, calibration to aquifer performance test data can be used in its place.

Step 6: Test the reliability of the model through a sensitivity analysis

The objective of a sensitivity analysis is to understand how different the solution could be, given a range of plausible parameter values. In short, the model calibration process has identified the best-fit values, and the sensitivity analysis summarizes the range of outcomes when one deviates from the best-fit values. A thorough model calibration is a sensitivity analysis. As such, the primary function of this step is to identify parameters that the model is particularly sensitive or insensitive to in the context of the project objectives. The sensitivity analysis is typically performed on the model calibration and documented in the calibration report, and it is then performed again on the scenarios simulated in step 7.

Step 7: Utilize the model to meet the objectives

Once the calibrated model is approved for use, the model is typically configured to run scenarios that will address the project objectives. The following steps are often taken when setting up and executing scenarios:

- Determine the background hydrologic conditions to be simulated. This can include the recharge, boundary fluxes, and pumping from areas that would be considered outside of the immediate area of scenario-specific interest.
- Determine the objective-specific conditions to be simulated and the range of values to be tested.
  - For example, if the objective was to estimate the hydraulic capture zones and the potential for saltwater intrusion resulting from 100, 200, and 300 gpm worth of pumping from a new water supply well situated near the coastline under average projected hydrologic conditions, three runs would be proposed with only the pumping rates at the proposed well varied.
  - If drought conditions were to be considered, six runs could be proposed using these three rates along with average projected conditions and drought conditions.
  - If sea level rise were also considered, additional scenarios would be included as required.
- Determine how the objectives are going to be assessed.
- Set up a file structure and nomenclature protocol to manage scenarios such that errors are avoided and archiving is effective.
- Perform and document the model simulations.

It is important for the utility staff and modeling team to agree on expectations related to this step early in the process (even in the proposal interview) to avoid conflict regarding scope creep and change orders related to the number of scenarios required.

Step 8: Documentation and long-term stewardship

The end products of the modeling effort are typically documentation of the model and the results as well as an archived version or versions of the model. The documentation will typically include plan view and/or cross-section graphics of (some or all of) simulated piezometric head contours, flow direction arrows, particle pathlines, concentration contours, and/or seawater interface position. The model documentation should include a tabulation of the water balance and a discussion of all of the steps included previously. Many of these features are included in the examples in Chapter 7.

The key model files, including those representing model calibration and the scenarios tested, should be delivered in an easy-to-understand and well-documented format, along with instructions on how to run the model, which software is needed, and, when applicable, how to install the software. Utilities can benefit by having staff trained on how to use the model for routine or potentially more complex tasks such as quarterly performance simulations or additional scenario testing.

A groundwater model can "live" for decades as an efficient tool for assessing present-day and yet-to-be-determined future questions. The capital invested in conceptualizing, building, calibrating, and delivering a model can become a useful tool for future aquifer assessment activities. It is therefore imperative that the model files be archived either at the client site, with the modeling team, or both (preferred) for potential future use.

Hypothetical Example of a Typical Groundwater Model Application, Model Objectives, and Approach

Anytown Water District has proposed to install eight new water supply wells within an aquifer in the presence of groundwater-fed streams, several groundwater contaminant plumes, and existing water users. A model will be developed to help estimate sustainable yield, prioritize well locations, delineate wellhead protection zones, and perform an environmental impact study. The model will incorporate future monthly pumping projections from both the eight new wells and existing regional pumping (80 wells total) within the model domain.

The model will be calibrated to transient historical conditions between the years of 1995 and 2015, which incorporate regional high and low water table conditions associated with wet and dry years, respectively. Monthly values of pumping, rainfall recharge, and boundary fluxes will be used as input for the model, which will be run using one-month time steps. Calibration to piezometric heads will be presented as residual plots for three synoptic rounds of water level measurements, corresponding to periods of wet, dry, and average conditions, and as hydrographs of simulated and observed piezometric head over time at 50 observation wells. Streamflow data from five stream gauges will be used to estimate groundwater baseflow and compared via hydrograph to simulated results at each location.

Transient (inputs and outputs vary with time) simulations of projected future conditions will be used to develop 30-year 3-D capture zones for each of the eight wells, as well as to estimate regional and local drawdowns and groundwater baseflow to streams. The future transient flowfield will incorporate seasonal and historic (cyclical) trends in recharge and pumping to evaluate the impacts of the eight new wells on a range of conditions. Since recent trends (the past five years) have indicated that drought conditions may be more pervasive in the future than they have been in the past, additional simulations will be run to understand the impacts on future operations should recent trends become the new operational "normal."

At the conclusion of the modeling activity, staff members will be trained on the model usage and continue to utilize the model periodically for capture-zone analyses. Model files will be archived at both the owner's and the modeling team's site.

USING THE RESULTS

Aquifers are areally extensive; as a result, the boundaries chosen are critical. A significant amount of money can be saved if some basic groundwater modeling based on the underlying governing equations is developed. Modeling should be performed before the full implementation of many water supply projects, especially where a more regional program with many wells is required. Computer models are designed to solve sets of these complex equations to demonstrate the ability of well fields to meet client needs. However, models model the average or average conditions; as such, results may be misleading if the geological characteristics of the aquifer system that are required to make the projects function properly in the long term (both technically and from a regulatory perspective) are lacking.

If the project is deemed feasible from technical, cost, and regulatory perspectives, a test program can be developed with appropriate monitoring wells and instruments installed. Once real data are obtained during the testing phase, the groundwater model can be recalibrated to assess the veracity of the initial assumptions and to make changes thereto. Therefore, understanding what the models mean is important in evaluating injection program effectiveness and success.

REFERENCES

Anderson, M.P.; Woessner, W.W.; and Hunt, H. 2015 (Vol. 4, 2nd ed.). Applied Groundwater Modeling: Simulation of Flow and Advective Transport. San Diego, Calif.: Academic Press.

Barnett B.; Townley, L.R.; Post, V.; Evans, R.E.; Hunt, R.J.; Peeters, L.; Richardson, S.; Werner, A.D.; Knapton, A.; and Boronkay, A. 2012. Australian Groundwater Modelling Guidelines: Waterlines Report. Canberra: National Water Commission.

Bloetscher, F.; Muniz., A.; and Witt, G.M. 2005. Groundwater Injection Programs. New York: McGraw-Hill.

McDonald, M.G. and Harbaugh, A.W. 1988. A Modular Three-Dimensional Finite-Difference Ground-Water Flow Model: US Geological Survey Techniques of Water Resources Investigations. Reston, Va.: USGS.

McDonald, M.G. and Harbaugh, A.W. 1984. A Modular Three-Dimensional Finite-Difference Ground-Water Flow Model. Reston, Va.: USGS (US Geological Survey).

Chapter 7

Case Studies Using Groundwater Models

OVERVIEW

This chapter builds on the content provided in Chapters 1–6 by providing a series of case studies in which groundwater models were used to address specific objectives of varying complexity for groundwater management purposes. There are many purposes to consider: quantity, quality, risk, and the combination of all of them. The simplest models may involve only one well with limited withdrawal (quantity). Other models may be concerned with safe withdrawals and the impacts of surface water bodies and land use on groundwater quality. The most complicated model may involve risk associated with injection projects to address public health concerns. The breadth of the case studies included here is intended to address many of these potential issues faced with respect to the use of models for groundwater management. Note that poorly constructed models may not identify what is really happening or modeling may not be conducted in a manner that identifies the actual impacts of pumping the groundwater. Models are averages of the averages, manipulated to find limitations of the water supplies. As a result, it should be expected that some models will *not* "find" water—i.e., certain situations may appropriately indicate that added or even current withdrawals are not sustainable.

The intent of the mix of case studies chosen is to highlight typical groundwater model usage and help provide a framework for answering the following questions as noted in Chapter 6:

- Will the use of a groundwater model add value to the project?
- What level of model complexity / effort is needed to satisfy the project objectives?
- Which type of model is needed to meet the project objectives?

- Are the data currently available sufficient to support the level of complexity required to answer the questions posed within a reasonable amount of uncertainty?
- How should a groundwater model be reviewed to ensure that the results are understood and that the recommendations made are based on best principles and a thorough understanding of the model's limitations?
- What factors should be considered when deciding how long-term stewardship of a groundwater model should be undertaken?

As noted in Chapter 6, a groundwater model attempts to "approximate reality" based on averaging conditions in the subsurface (the averages of the averages) while being transparent regarding its uncertainties and limitations over a potentially open-ended time frame. As such, a groundwater model can "live" for decades as an efficient tool for assessing present-day and yet-to-be-determined future questions.

SUMMARY OF CASE STUDIES

While countless groundwater models have been developed over the decades to answer a wide array of questions, for this manual, the authors selected these models to demonstrate the following points. Simple algorithms can serve as a type of model for questions that only require basic approximations for understanding the problem and possible solutions. The characteristics of groundwater—such as recharge, discharge, flow direction and movement through changes in the subsurface rock strata, and migration of constituents—should be addressed specific to the aquifer and to the question(s) being addressed by the model. Therefore, models were selected from around the United States. The case studies span a range of complexity, geographic region, model size, and objectives. Each is briefly summarized as follows (with expanded versions found in the next section of this chapter). Note that each of these models add complexity over the prior model as a means toward understanding groundwater (Table 7-1).

The Saco River model is a straightforward groundwater well model using MODFLOW, a US Geological Survey (USGS) modeling platform. The Iowa model looked at recharge that might include nitrates in the water. The Floridan model is a confined saltwater system, while the North Dakota model was impacted by Prior Appropriation Doctrine water rights. The California and Suffolk County models are far larger areas and look at surface interfaces, contaminant transport, and sea level rise. The Hollywood model is by far the most complex and asks the most poignant question about recharge and public health and, therefore, needs the most discussion. It looks at the risks of groundwater recharge with reclaimed wastewater, an indirect potable reuse concept that more than a dozen utilities are exploring or have explored. The model includes saltwater density models and risk contouring in ArcGIS. The goal of these models was to provide useful information to inform decision-making and provide insight on what can be accomplished.

There are many cases where there are issues involved with modeling that fail to address the larger need. Southwestern Utah is an area where the state required groundwater models, but the aquifer continued to fall at a fast rate. Other areas have been slow to model; populated areas of Arizona, much of the Snake Valley, the Ogallala, aquifers in the Great Lakes, and the Carolinas are examples where aquifer levels fell dramatically before efforts were made to address water supply concern.

Table 7-1 Key features of case study models

Model Subject/ Area	Owner/ Location	Reference	Model Approach and Complexity	Topic Addressed	Number of Wells
1. Saco River Valley glacial groundwater	State of New Hampshire	Tepper et al. 1990	2-D, steady-state, finite difference model	Groundwater flow and particle tracking	>10
2. Induced surface water recharge	Iowa Deptartment of Natural Resources, Dunlap, Iowa	Langel and Gannon 2011	3-D numerical, transient conditions, finite difference model	Evaluate quantity and quality of induced recharge	2
3. Assessment of public water supply wells	Upper Floridan Aquifer, Fla.	Jagucki et al. 2009	3-D steady-state, finite difference model	Assess contaminant risk to public wells	77
4. Groundwater supply evaluation	Stutsman Rural Water District (SRWD), N.D.	SRWD application to NDSWC 2013	Transient with variable time steps, finite difference model	Evaluate groundwater resource and water appropriation permit	18
5. Platte West Well Field	Metropolitan Utilities District, Omaha, Neb.	City of Omaha 2005, 2012	2-D and 3-D, steady-state and transient, finite difference model	Evaluate groundwater pumping, induced infiltration from surface water	42
6. Central Valley Groundwater-Surface Water Simulation, with contaminant transport	California Deptartment of Water Resources (DWR), Central Valley, Calif.	California DWR, Brush and Dogrul 2013	3-D, transient with variable time steps, multiple years, finite element model	Use of integrated groundwater–surface water flow model to assess sustainable management, drought planning, wellhead protection, land use	21 model subregions
7. Suffolk County Water Management Program including sea level rise	Suffolk County Deptartment of Health, N.Y. (Long Island)	Suffolk County website	3-D, transient with variable time steps over multiple years, finite element model	Groundwater flow paths, contaminant transport, saltwater movement, sea level rise, baseflow to streams	>700
8. Salinity barrier	City of Hollywood, Fla.	Bloetscher et al. 2019, Phonpornwithoon 2016	Simulation of variable-density flow and groundwater transport	Evaluate hydraulic barrier to saltwater intrusion and quantity of available groundwater	1

Source: Created by author

DETAILED CASE STUDIES

Case Study 1: Saco River Valley Glacial Aquifer Groundwater Model

The 2-D numerical model that was used for the Saco River project was developed more than 25 years ago by USGS and has been used to support water supply decision-making in the years since. This example is included to illustrate the value of adopting existing USGS models for additional uses and to meet project objectives at a lower level of effort.

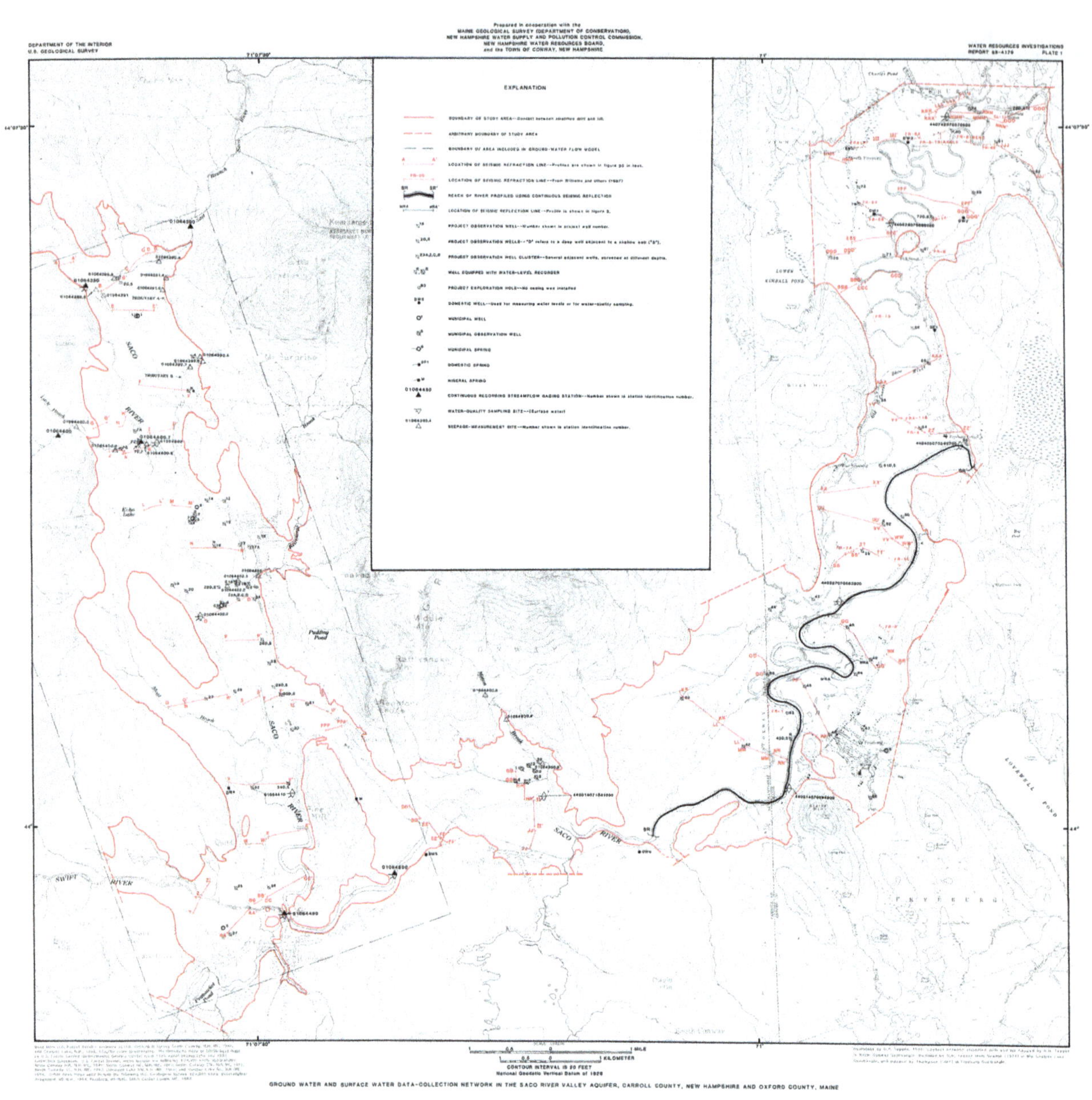

Source: Tepper et al. 1990

Figure 7-1 Saco River Valley

As indicated by Tepper et al. (1990), the Saco River Valley is located in the foothills of the White Mountains of New Hampshire (see Figure 7-1). The valley contains an unconfined glacial aquifer that ranges in width from 1 to 3 mi. It consists primarily of unconsolidated sand and gravel deposits bounded by clay, till, or bedrock on the sides and bottom. The principal flow direction in the aquifer is from the till-covered or bedrock uplands toward the Saco River. The aquifer is relatively thick and contains high-transmissivity zones along the river, where pumping yields are good to excellent.

USGS developed and calibrated a numerical groundwater flow model of the Saco River Valley as presented in Water Resources Investigations Report (WRIR) 88-4179 (Tepper et al. 1990) using the MODFLOW simulation program. This work was performed using cooperative funding from the State of New Hampshire, with the express purpose of using the results for water supply planning and protection. The active model domain covers an area of approximately 15 mi^2 and extends from the junction of the Saco and Ellis Rivers to the junction of the Swift and Saco Rivers near Conway Village (as shown in Figure 7-1 excerpted from WRIR 88-4179). The model uses a block-centered finite difference grid containing 1 layer, 68 rows, and 162 columns with uniform spacing of 300 ft on a side.

The model was calibrated to long-term water-level conditions using recharge from both precipitation and runoff from upland areas (Tepper et al. 1990). The calibrated model was used to predict the effects of increased pumpage under varying recharge conditions on groundwater levels and on the size of contributing areas to municipal wells.

Since its creation in 1990, this 2-D model has been used by water supply utilities to estimate wellhead protection zones, delineate zones of capture, and assist with the permitting and siting of proposed wells. Since this model is publicly available, limited additional resources need to be spent developing, calibrating, and validating the model. Where simple analyses are needed, the use of an existing USGS groundwater model can be an efficient choice, even if it is over 30 years old.

Since the mid-1990s, all USGS models have been archived in the USGS water science center that built the model(s). All USGS models completed since 2010 are also publicly archived and available for download from the USGS Science Data Catalog using the keyword "usgsgroundwatermodel". The USGS publications warehouse is another tool that can help a utility determine what studies and models have been completed in an area of interest.

Project name: Saco River Valley Glacial Aquifer Groundwater Model

Agency: USGS

Location: New Hampshire

Project objectives: Examine the impacts of pumping on the aquifer

Size of model domain: 15 mi^2

Dimensions: 2-D

Temporal coverage: Steady-state

Computations: Finite difference

Processes simulated: Groundwater flow, particle tracking

Number of drinking water wells included: >10

Software: MODFLOW

Case Study 2: Numerical Modeling to Evaluate Quantity and Quality of Induced Surface Water Recharge

This 3-D model developed by the Iowa Department of Natural Resources includes surface water interaction with the primary drinking water aquifer and two City of Dunlap water supply wells. Capture zones were delineated to guide management decisions with respect to land use measures such as fertilizer application. This example is included as a moderate-complexity model application in which surface water is explicitly incorporated.

Langel and Gannon (2011) completed a hydrogeologic study of the Boyer River alluvial aquifer near Dunlap, Iowa, in 2010 for the Iowa Department of Natural Resources. The objectives of the study were to use a numerical groundwater model to quantify the amount of induced recharge received from the Boyer River by two public supply wells and to delineate the capture zones of those two wells. The results of the study were used to direct land use management to reduce nitrate levels around the City of Dunlap's well field.

The three-layer groundwater flow model was developed in MODFLOW, covered an area of approximately 5.25 mi^2 (Figure 7-2), and included the Boyer River alluvial aquifer; the overlying silt and clay layer; the underlying till, shale, or sandstone layer; and the Boyer River itself (Figure 7-3).

The model was then calibrated to steady-state and transient conditions, and particle tracking analyses were completed using MODPATH. Particles were released from the two City of Dunlap public supply wells (pumped at a combined rate of 560 gpm) and tracked backward in space and time to their recharge locations. The two- and five-year capture zones are shown in Figure 7-4 and cover 82 and 106 acres, respectively. By examining the recharge locations of the capture zones, land use management measures could be proposed to minimize the impact of nitrate on the supply wells. One example could be reducing the application of nitrogen fertilizer on the sports complexes or agricultural land located within the capture zones.

The city wells were not found to receive induced recharge from the Boyer River. However, induced recharge is possible during short-term durations when the elevation of the Boyer River is higher. Locating new wells closer to the Boyer River may improve the water quality by capturing more induced recharge.

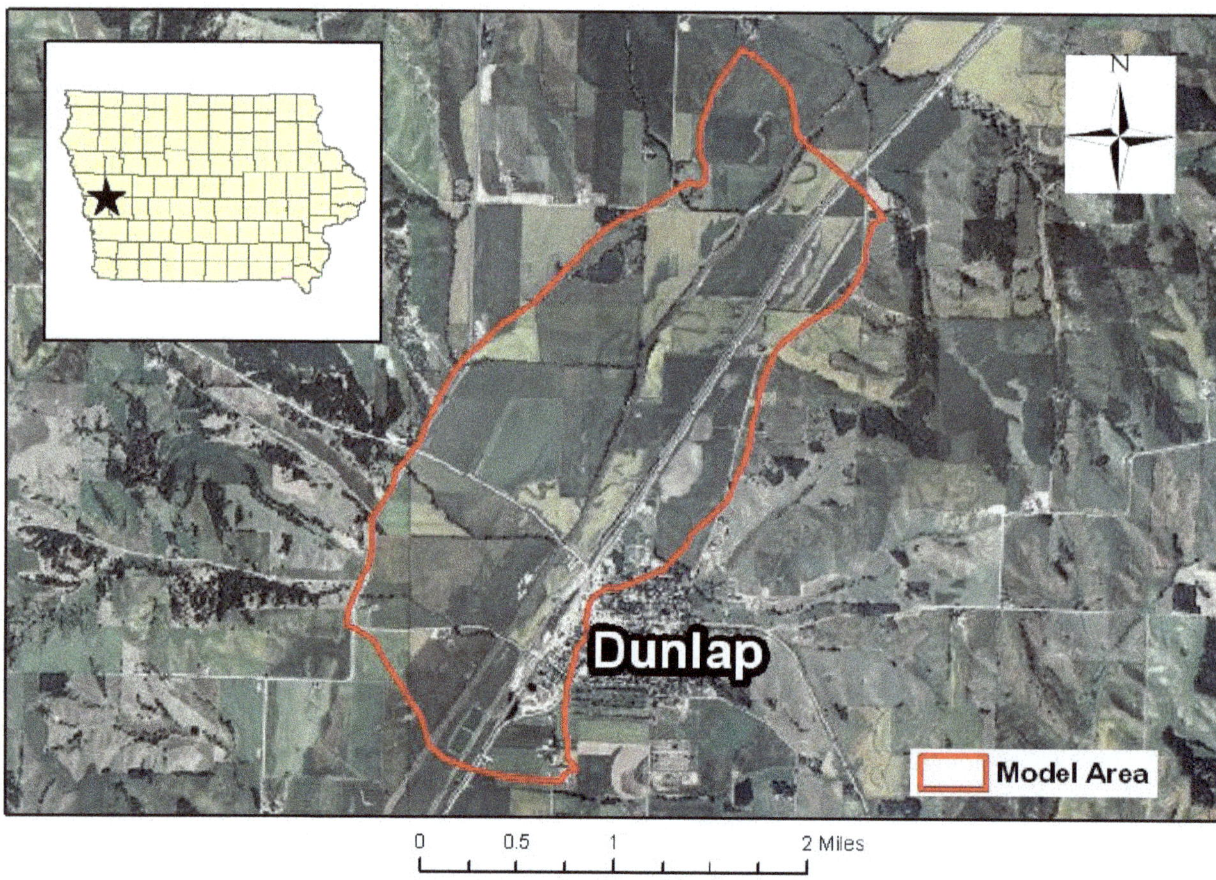

Source: Langel and Gannon 2011

Figure 7-2 Three-layer groundwater flow model developed in MODFLOW

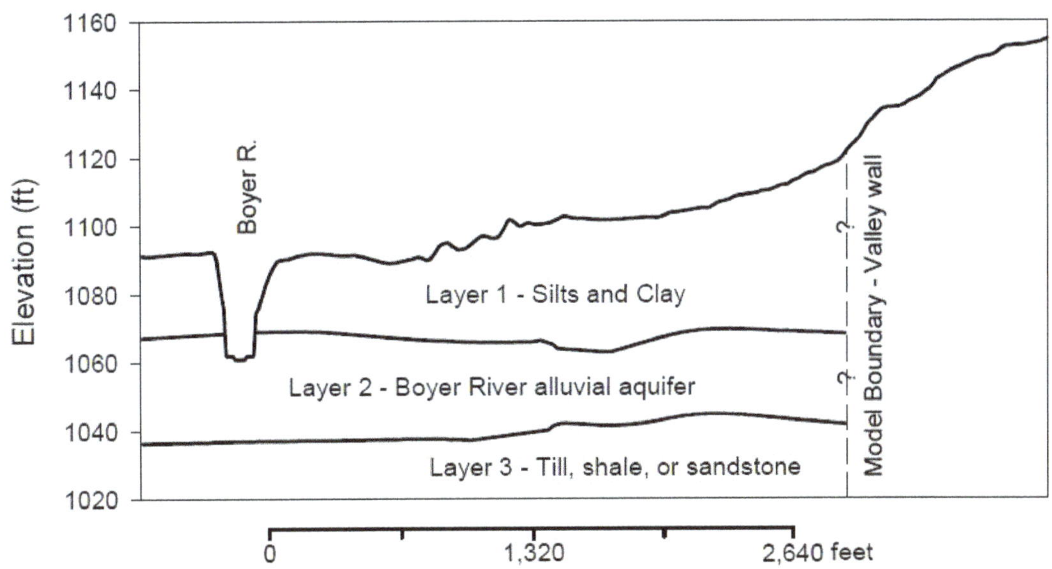

Source: Langel and Gannon 2011

Figure 7-3 Aquifer formation profile

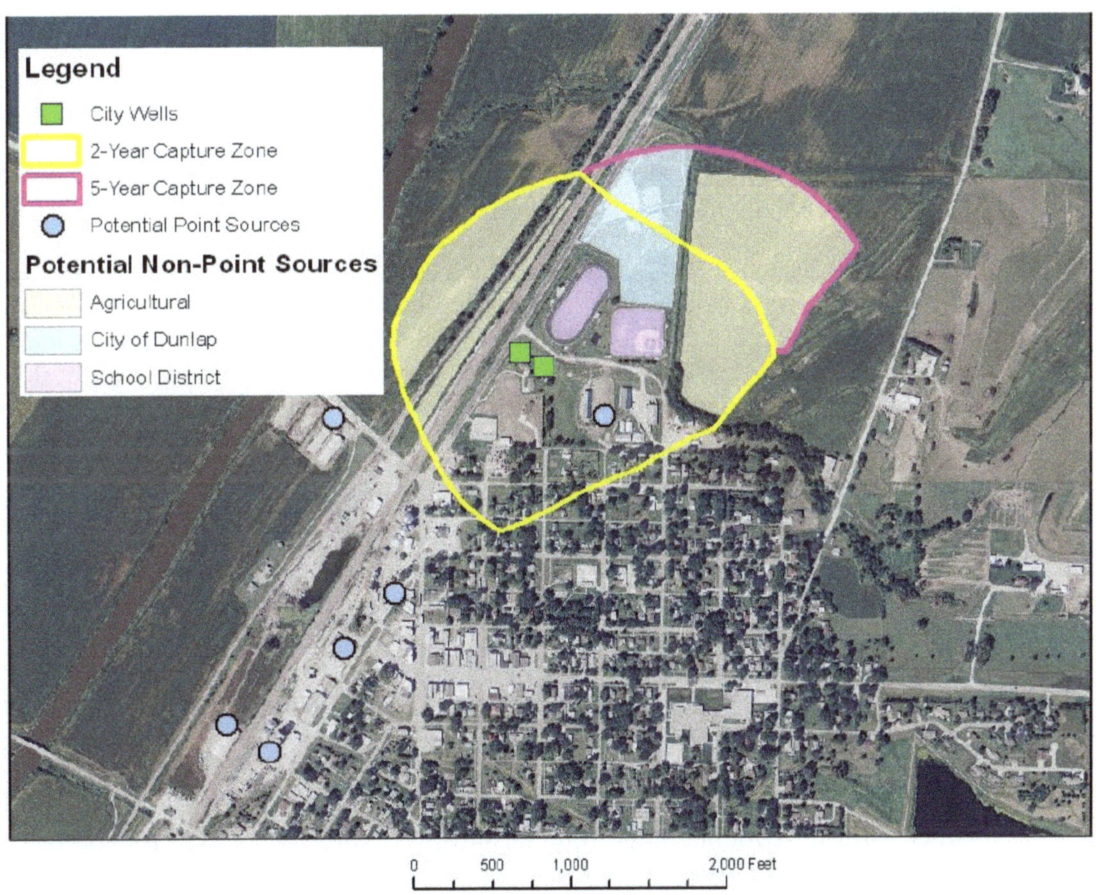

Source: Langel and Gannon 2011

Figure 7-4 Two- and five-year capture zones for Dunlap well field

Project name: Numerical Modeling to Evaluate Quantity and Quality of Induced Surface Water Recharge

Agency: Iowa Department of Natural Resources

Location: Dunlap, Iowa

Project objectives: Model-induced surface water recharge volume and quality

Size of model domain: Approximately 5.25 mi^2

Dimensions: 3-D

Temporal coverage: Transient

Computations: Finite difference

Processes simulated: Groundwater flow, particle tracking, surface water interaction

Number of drinking water wells included: 2

Software: MODFLOW, MODPATH

Case Study 3: Assessing the Vulnerability of Public-Supply Wells to Contamination – Upper Floridan Aquifer System

This 3-D model developed by USGS was used to understand the capture zone and age of groundwater extracted from a public supply well in the Upper Floridan aquifer near Tampa, Fla. This example was included to illustrate the complexity and uncertainty involved when modeling fractured rock, such as the limestone of the Upper Floridan aquifer. A heavy reliance on data interpolation and parameter estimation (this example uses a separate code, UCODE, to estimate parameter values) is required to represent the subsurface under these conditions.

USGS, through its National Water-Quality Assessment Program, performed a detailed study of the vulnerability of a public supply well in Temple Terrace, Fla., northeast of Tampa. The well typically produces water at the rate of 700 gpm from the Upper Floridan aquifer, which is carbonate rock, and is overlain by a surficial sand unit with varying amounts of silt and clay referred to as the surficial aquifer system and an intermediate depth confining unit. The thickness of the intermediate confining unit is variable throughout the area and contains numerous breaches through which groundwater can move rapidly downward into the Upper Floridan aquifer. Groundwater in the Upper Floridan aquifer may move slowly through the porous matrix or may travel at a relatively rapid rate through the many fractures and cavities in the limestone as depicted schematically in Figure 7-5.

A local-scale, steady-state (to year 2000 conditions), 3-D groundwater flow model was used to estimate the zone of contribution to the public supply well (Crandall et al. 2009; see Figure 7-6). The model used MODFLOW, UCODE, and MODPATH. UCODE is a parameter-estimation program that can be used with any program with numerical input and output files to refine parameter (such as hydraulic conductivity and porosity) estimates. The model covers an area of 7.5 mi^2 and uses 13 computational layers to represent six hydrologic units. Model layers 5–11 represent the Upper Floridan aquifer, and model layer 8 represents the highly transmissive cavernous zone set to a constant thickness of 3 m throughout the model. This zone was approximated based on caliper and video logs of the public supply well. Additional karst features are incorporated into the model by using larger vertical hydraulic properties, anisotropy, high recharge, and low porosity values where karst features are thought to be present. Preferential groundwater flow direction because of fracture orientation is simulated in the model by specifying anisotropy.

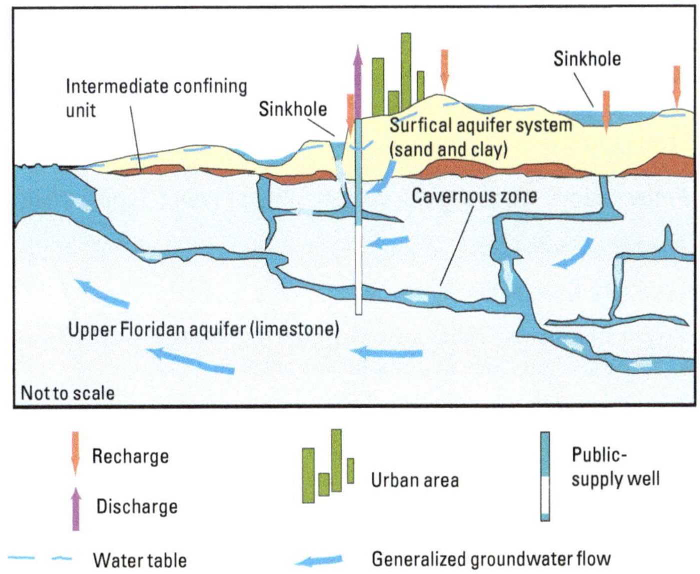

Source: Jagucki et al. 2009

Figure 7-5 Upper Floridan aquifer under the Tampa, Fla., area

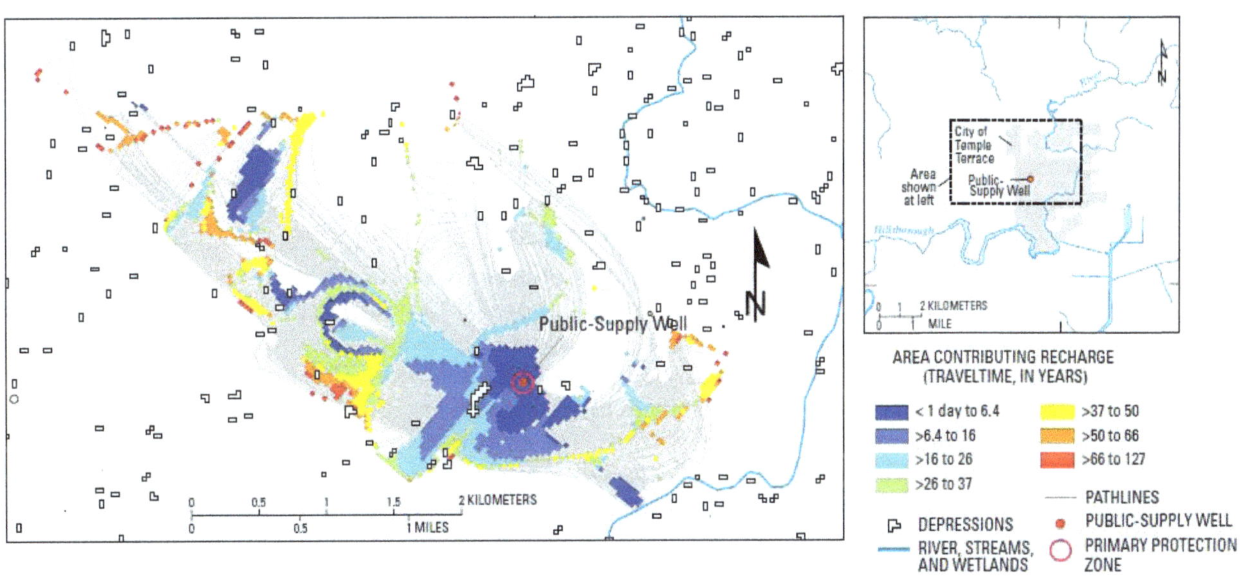

Source: Jagucki et al. 2009

Figure 7-6 Simulated area of recharge

Particle tracking was used to compute flow paths and advective travel times in the model area. The computed area contributing recharge (ACR) for the public supply well covered about 9.3 mi^2, and the simulated ages of the particles reaching the public supply well ranged from 1 day to 127 years with a median of 13.1 years. Groundwater older than about 80 years within the ACR would be expected to contain primarily agricultural or related contaminants, whereas groundwater younger than 80 years would be expected to contain both urban and agricultural contaminants.

A key finding of this case study was that the youngest water entering the well does not recharge the aquifer near the well field (Figure 7-6). Therefore, restricting land use activities in a 500-ft radius around the well (or other wells in similar settings) would not protect the well(s) from human activities in many areas associated with rapid groundwater travel times to the well.

Project name: Assessing the Vulnerability of Public Supply Wells

Agency: USGS

Location: Tampa, Fla., area

Project objectives: Delineate contributing recharge areas and water age to characterize contaminant risk at a public supply well

Size of model domain: Approximately 86.24 km^2

Dimensions: 3-D

Temporal coverage: Steady-state

Computations: Finite difference

Processes simulated: Groundwater flow, particle tracking

Number of drinking water wells included: 77

Software: MODFLOW, MODPATH, UCODE

Case Study 4: Stutsman Rural Water District Groundwater Supply Hydrogeologic Evaluation

This 3-D model was included to illustrate an example of how a groundwater model can be used to assess impacts within rural areas governed by the Doctrine of Prior Appropriation. The Stutsman Rural Water District (SRWD) is a rural water utility that provides drinking water to approximately 6,000 people, including individual service connections and three bulk water connections. SRWD also provides water for industrial use to local industries. In 2013, SRWD submitted a water appropriation application to the North Dakota State Water Commission (NDSWC) requesting approval to pump approximately 4,800 ac-ft/yr (approximately 3,000 gpm) from the Spiritwood aquifer for a fertilizer plant. A numerical groundwater computer model was developed to determine if the requested groundwater appropriation could be pumped from the Spiritwood aquifer without adversely impacting existing industrial, public water supply, and irrigation wells (prior appropriators).

The Spiritwood aquifer is composed of sand and gravel deposited by meltwaters associated with continental glaciation experienced by the northern Great Plains during the Pleistocene epoch. The sand and gravel outwash deposits are buried by 50–100 ft of glacial drift. The glacial drift is composed primarily of till, which is a heterogeneous mixture of mostly clay but also includes silt, sand, gravel, cobbles, and boulders. The Spiritwood aquifer consists of a series of interwoven or braided buried valleys or channels extending from Canada to South Dakota, and the project area is located near Jamestown, N.D.

The Spiritwood aquifer is a confined aquifer, with the potentiometric surface (the height the water would rise in a cased well) as much as 100 ft above the top of the aquifer. The aquifer properties of transmissivity, storativity, and hydraulic conductivity were determined both from literature and from a constant-rate aquifer pumping test conducted using a new well constructed by SRWD at the proposed diversion point. The aquifer pumping test consisted of pumping the well at the rate of 2,300 gpm for seven days and monitoring water levels (before, during, and after the pumping period of the test) in 43 observation wells (most of which were part of the NDSWC Spiritwood aquifer monitoring network).

Water appropriation in North Dakota is by the Doctrine of Prior Appropriation. This water appropriation system is often identified by the phrase "first in time, first in right." Each permitted water use is assigned a priority date. In those instances where there is a limited amount of groundwater available—because of scarcity of the resource, high demand for the resource, or both—the most senior water right holders' permitted amounts are protected. To receive a water appropriation permit, the applicant must meet four criteria: the rights of a senior water right holder cannot be unduly affected, the proposed means of diversion must be adequate, the proposed use of water is beneficial, and the proposed use is in the public interest. The proposed industrial use of the water for which SRWD was seeking a permit met criteria 2, 3, and 4. The primary purpose of the groundwater model was to determine if the requested appropriation would meet criterion 1.

A hydrogeologic conceptual model (HCM) was developed to identify the hydrogeologic units, hydrologic processes, and hydrologic boundaries that control groundwater flow in the project area. While constructing the HCM, hundreds of geologic logs maintained by North Dakota state agencies in two online databases were reviewed, and the information was used to refine the aquifer delineation. These databases revealed that portions of the aquifer are cross-cut by low-permeability (LP) barriers. These LP barriers are areas where the aquifer material was either not deposited or where the aquifer material is rich in fine-grained silts and clays. These LP barriers, some of which were previously identified (Shaver and Pusc 1992) and some that were identified during the groundwater supply hydrogeologic evaluation, are both longitudinal and transverse relative to the buried channel. The LP barriers restrict groundwater flow to variable degrees and have a significant impact on groundwater movement in the aquifer.

A key consideration of the HCM and an important model input parameter is aquifer recharge. A significant source of recharge to the aquifer was thought to be infiltration of water from the overlying glacial drift, particularly near the pumping centers in the aquifer. In a previous Spiritwood aquifer investigation conducted in a different location, it was reported that irrigation withdrawals are primarily derived from leakage through the overlying glacial drift and not from storage in the Spiritwood aquifer (Shaver 1994). To confirm that pumping-induced recharge would occur during the proposed SRWD withdrawals under the new permit, nested observation wells completed within the aquifer and the glacial drift (Figure 7-7) were constructed and monitored during the aquifer test. During the aquifer test, drawdown induced by pumping was documented in an observation well completed in the overlying drift, thereby confirming that pumping-induced aquifer recharge from the overlying drift was possible.

Recharge and ET were applied to each stress period using recharge and ET data estimated by the NDSWC utilizing the Versatile Soil Moisture Budget Model (VB 2000). VB2000 was developed for use on the Canadian prairies to estimate available water within the root zone (Honeyman 2014). The annual recharge rate used in the transient model ranged from a low of 0.51 in./yr in 1976 to a high of 5.51 in./yr in 1994. The recharge analysis, and subsequent use in the model, was complicated by climatic variations experienced in the region during the past 43 years. Relatively abundant precipitation in the project area from approximately 1993 to 2013 (wet period) resulted in recharge rates that ranged from 3 in./yr to more than 5 in./yr. During the periods of relatively low precipitation (dry period) from 1970 to 1989, the resulting recharge rate ranged from less than 1 in./yr to nearly 2 in./yr. The average annual recharge rate over the entire period evaluated for the model was 1.7 in./yr.

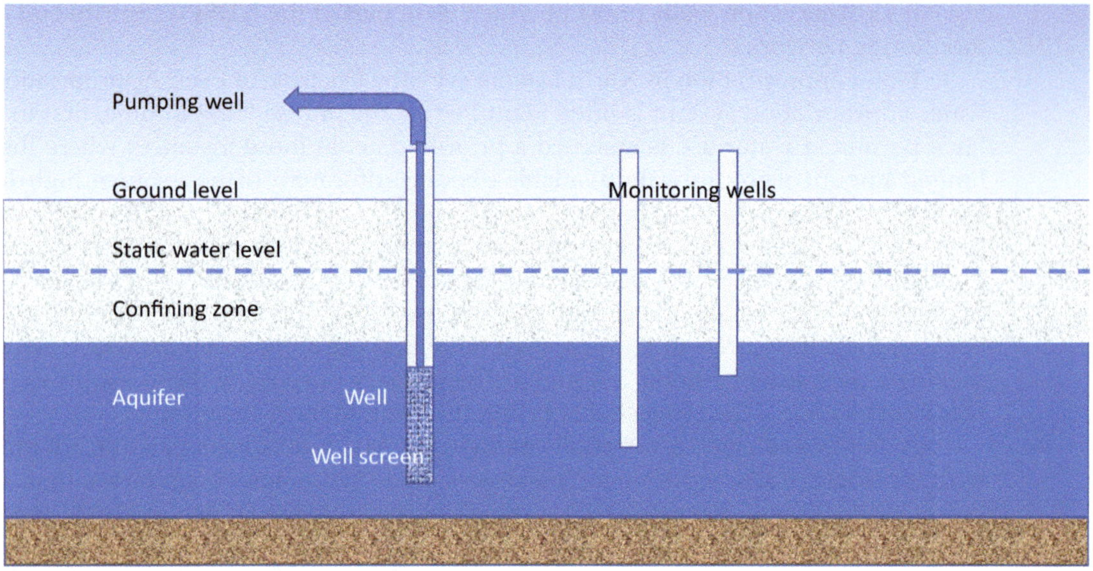

Source: Created by author

Figure 7-7 Nested well schematic, showing pumping-induced recharge

Two transient model calibrations were conducted—one based on long-term water level data and one based on the short-term changes in water levels measured during the aquifer test conducted utilizing the well at the proposed diversion point. The long-term transient model calibration was performed by comparing model-calculated water levels to the water levels measured in approximately 70 NDSWC observation wells in the project area. The transient model calibration was performed over the period of 1974–2012, and the aquifer and glacial drift hydraulic properties were iteratively refined through the transient calibration process.

Steady-state simulations using the calibrated model were used to predict the aquifer response to added withdrawals associated with the proposed new industrial use, particularly relative to water level drawdown at wells used by senior water permit holders. Simulations were conducted using recharge rates based on the current relatively wet conditions, as well as recharge rates based on the dry periods in the historic record. The predictive model simulation results indicated that withdrawals of up to 1,000 gpm from the proposed diversion point, under current pumping and climatic conditions (recharge rate likely greater than 3 in./yr), would not likely cause adverse impacts to senior water permit holders in the project area (excluding two irrigation wells, which had not been pumped for years because of the wet conditions and were expected to be mitigated by administrative measures).

In the end, the proposed fertilizer plant was not constructed (because of a variety of reasons), making the requested appropriation amount of 4,800 ac-ft/yr no longer necessary. However, added withdrawals by other industrial users in the future are likely, and the results of the model and related hydrogeologic evaluation provide an additional tool with which to assess the potential impacts of increased withdrawals from the aquifer on senior water rights holders.

Project Name: Stutsman Rural Water District Groundwater Supply Hydrogeologic Evaluation

Stakeholders: Stutsman Rural Water District, Jamestown, N.D.; North Dakota State Water Commission, Bismarck, N.D.

Location: Stutsman and Barnes Counties, N.D.

Project objectives: Groundwater resource evaluation, water appropriation permit analysis

Size of model domain: Approximately 85,000 ac

Dimensions: Three

Temporal coverage: Transient, variable time steps, multiple years

Computations: Finite difference

Processes simulated: Groundwater withdrawals, groundwater recharge

Number of wells included: 18 (public water supply, industrial and irrigation)

Software: GMS by Aquaveo and Versatile Soil Moisture Budget VB2000

Case Study 5: Metropolitan Utilities District, Platte West Wellfield

This 3-D model quantifies the impacts of the inclusion of 42 new water supply wells designed to provide an average of 52 mgd to the Greater Omaha area. Of particular concern to the Omaha Metropolitan Utilities District (District) was the continued containment of contaminated groundwater at a Superfund site approximately 3 mi west of the proposed wells. This example was included to illustrate an environmental impact assessment involving a significant change to an aquifer that contains a groundwater contamination plume.

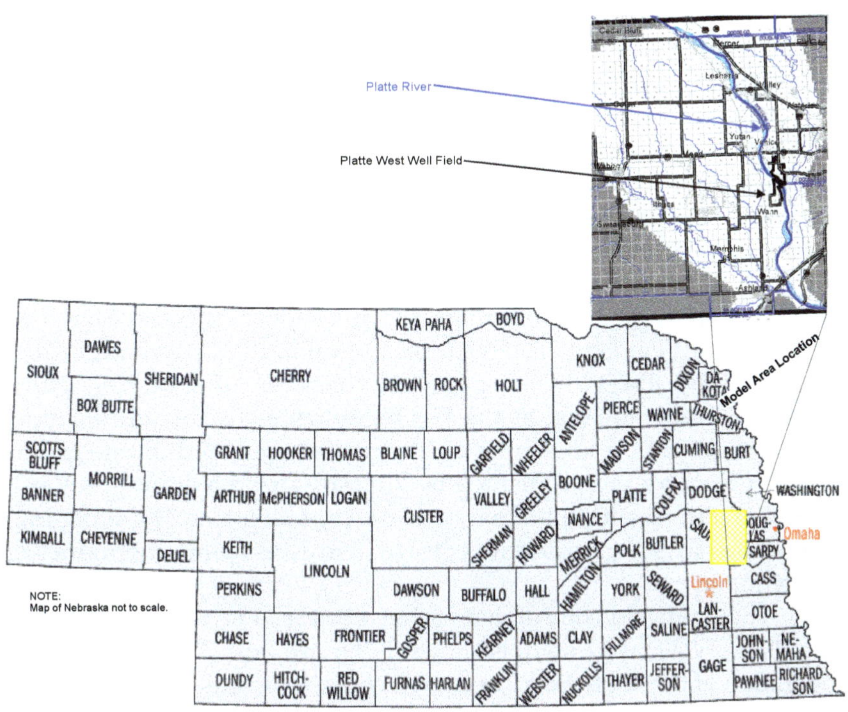

Source: Omaha Metropolitan Utilities District 2012

Figure 7-8 Location of the Platte West wellfield

The District provides drinking water to more than 210,000 customers in the Greater Omaha, Neb., area. Around 2004, the District began planning a new well field, the Platte West well field (Platte WWF). The Platte WWF is designed to provide an average of 52 mgd to the Greater Omaha area from 42 production wells completed in the alluvium of the Platte River valley. The Platte WWF was brought online in 2009, and withdrawals from the well field have generally ranged from 28 to 36 mgd. The location of the Platte WWF is shown in Figure 7-8.

A Superfund site, the former Nebraska Ordinance Plant (FNOP), is located approximately 3 mi west of the southwest boundary of the Platte WWF. The past disposal of radioactive and other solid wastes at the FNOP contaminated the shallow groundwater, and remedial wells are in operation at the FNOP for the purpose of controlling and preventing the migration of contaminated groundwater from the FNOP (hydraulic containment). Another relatively large well field, the Ashland well field, is operated by the City of Lincoln, Neb. The Ashland well field also pumps groundwater from the Platte River alluvium, located approximately 5 mi south of the southern boundary of the Platte WWF. During the planning and design of the Platte WWF, a groundwater model was developed to assess the effect that operation of the Platte WWF would have on the FNOP hydraulic containment system and on the configuration of the FNOP contaminated groundwater plume, as well as to assess what effect the operation of the Platte WWF would have on water levels in the Ashland well field.

The model domain is shown in Figure 7-9. The east and west no-flow boundaries are the east and west edges of the alluvial aquifer. The northern no-flow boundary is a groundwater divide, formed by the ancestral channel of the Platte River. The south boundary includes river cells, and there are river cells representing six streams in the model domain. Also included in the model domain are drain boundary cells that simulate the removal of groundwater by agricultural drain tiles. The model has one layer representing the unconsolidated alluvial sediments. Underlying the unconsolidated alluvial sediments is consolidated bedrock (shale or sandstone). The bedrock units have low hydraulic conductivity, and there are little to no groundwater withdrawals from the consolidated bedrock units underlying the alluvium. Consequently, the base of the model is a no-flow boundary.

Data used in the construction of the model included geologic logs from the Nebraska Department of Natural Resources groundwater well database, geologic logs from the FNOP Superfund site, and geologic logs from test drilling conducted during the Platte WWF design. The model simulates the aquifer from 2004 to 2015. Data used to calibrate the model included streamflow and stage data obtained from six USGS stream-gauging stations and groundwater elevation data from up to 190 area monitoring wells.

Model calibration included steady-state calibration to groundwater elevations measured in the area monitoring wells in October 2003 and March 2005 (before bringing the Platte WWF online in 2009). A second calibration approach included particle tracking utilizing MODPATH. This approach consisted of inputting particles at the known source of contamination at the FNOP, utilizing the model to simulate the transport of the contaminants, and comparing the model-simulated plume geometry with the actual plume geometry as documented by groundwater monitoring. A third calibration method included a transient calibration to a 96-h pumping test. The pumping test was conducted at the rate of 2,300 gpm and was designed to assess pumping-induced infiltration from the Platte River.

The predicted cone of depression, produced by the model-simulated pumping from the Platte WWF at 52 mgd, is shown in Figure 7-10. The model results indicated that the operation of the Platte WWF at design capacity would not impact the efficacy of the FNOP hydraulic containment system nor would it alter the travel path of the contaminant plumes. The model results also indicated that 71 percent of the water pumped from the Platte WWF would be derived from direct induced surface water infiltration from the Platte River (see Figure 7-10).

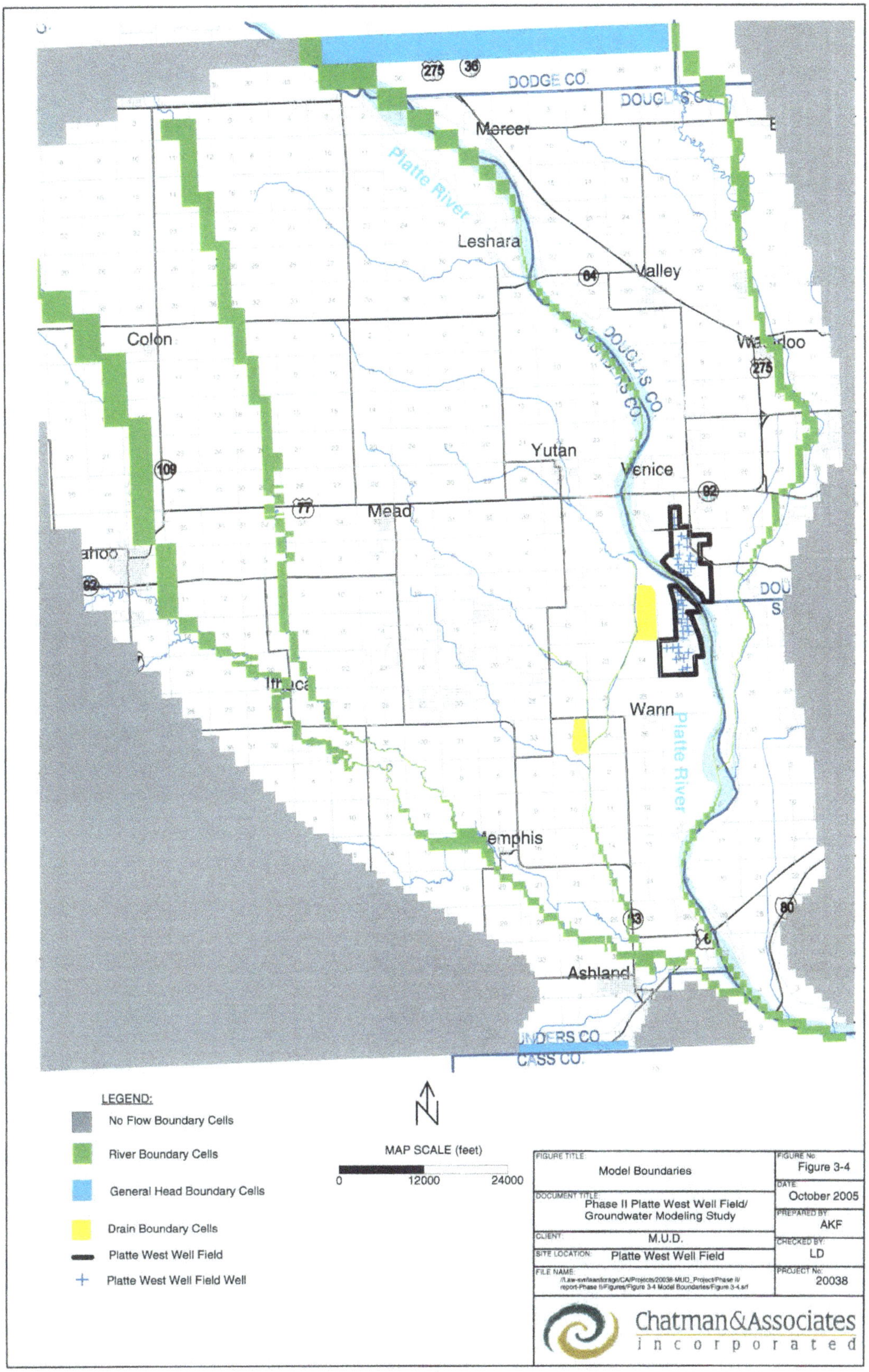

Source: Omaha Metropolitan Utilities District 2005

Figure 7-9 Groundwater model domain/boundaries for Platte West wellfield (WWF) project

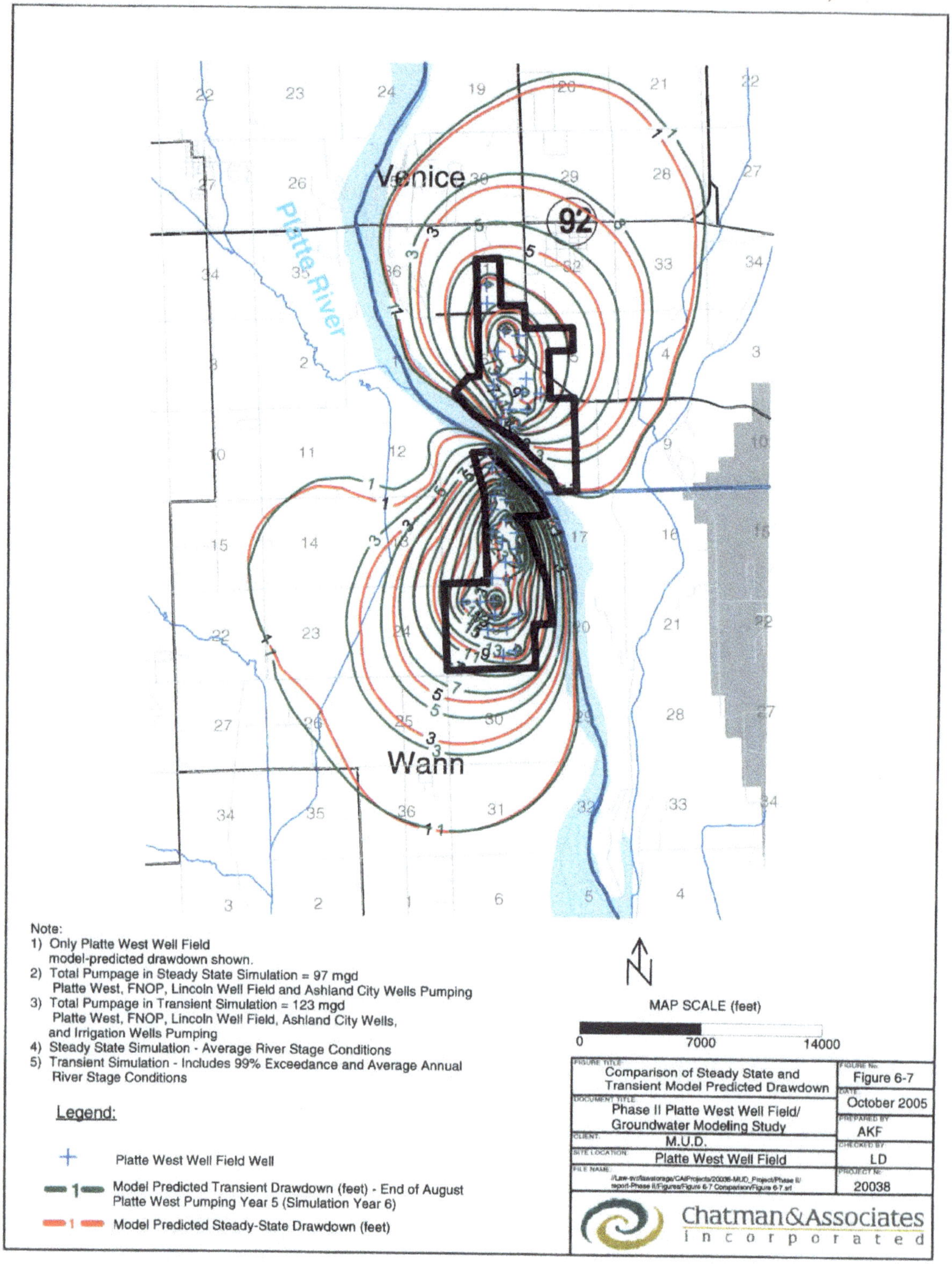

Source: Omaha Metropolitan Utilities District 2005

Figure 7-10 Model-simulated pumping from the Platte West wellfield at 52 mgd

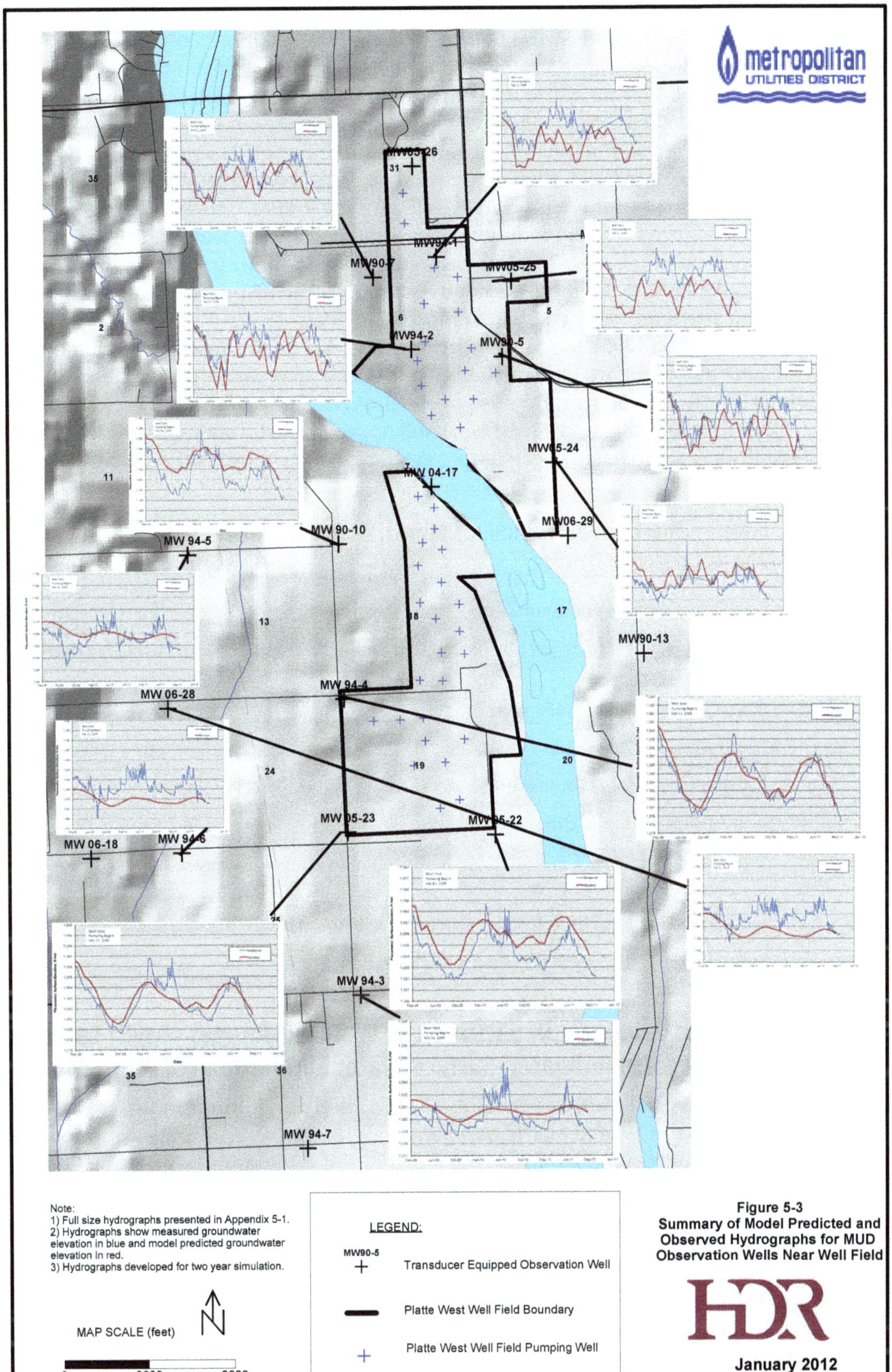

Source: Omaha Metropolitan Utilities District 2012

Figure 7-11 Calibration of predicted versus actual results from modeling

Because the Platte WWF is an operating wellfield, the utility performs annual reviews of the results measured in the field versus the model since 2009. These annual reviews provide the opportunity to refine the model, thereby increasing the model's utility as a predictive tool. The annual reviews include the following: inputting the monthly average production rate of the 42 Platte WWF wells, inputting the average monthly stage for each of the six streams in the model domain, and conducting transient simulations and comparing the model-calculated water levels to actual measured water levels in 19 monitoring wells. The results for 14 of the monitoring wells, for 2009–2012, are shown in Figure 7-11. The graphs in Figure 7-11 illustrate good agreement between water levels predicted by the model versus actual water levels measured in the monitoring wells. The annual postaudit results indicate that the model accurately predicts the response of the aquifer to the new pumping stress represented by operation of the Platte WWF.

The original 2004 model predictions have been confirmed with actual monitoring well data, indicating that the model is a reasonably accurate representation of the alluvial aquifer in the project area. Because of the proven ability of the model to reproduce real-world conditions, the model has been used to provide information regarding source water classification and has also been used to evaluate and optimize the Platte WWF pumping schedule.

Project name: Metropolitan Utilities District, Platte West Well Field

Stakeholders: Omaha Metropolitan Utilities District, Omaha, Neb.

Location: Douglas and Saunders Counties, Neb.

Project objectives: Sourcewater protection, well field management, surface water impact assessment

Size of model domain: 650 mi^2

Dimensions: 2-D and 3-D

Temporal coverage: Transient and steady-state, annual model updates

Computations: Finite difference

Processes simulated: Groundwater withdrawals, induced infiltration from surface water, contaminant transport

Number of wells included: 42 production wells, >100 observation wells

Software: MODFLOW 2000 and MODPATH

Cast Study 6: California Department of Water Resources (DWR) California Central Valley Groundwater–Surface Water Simulation (C2VSim) Model

California's Central Valley is the hub of the state's agricultural industry and grew from 732,000 people in 1920 to nearly 8 million people in 2010, with projected growth to more than 11 million people by 2050. During this time, the annual value of Central Valley agricultural products has grown to more than $21 billion. The Central Valley is also the hub of the state's water supply system; an extensive network of dams and canals supplies surface water to users within the Central Valley and in the San Francisco Bay Area, Central Coast, and Southern California. Agricultural and urban users within the Central Valley consumed on average 13 million ac-ft of surface water and more than 8 million ac-ft of groundwater per year between 2000 and 2009, and up to an additional 2 million ac-ft of water was exported to areas outside the Central Valley.

The availability of surface water supplies in California varies significantly from year to year, and several dry years can result in critically low water reserves. The availability of surface water supplies is also constrained by regulatory restrictions including downstream water rights, in-stream flow requirements on select river reaches, and outflow requirements for the Sacramento–San Joaquin Delta. Groundwater is often used as a buffer when surface water supplies are reduced, but over-reliance on groundwater can have serious negative consequences including declining water levels, reduced water quality, and land subsidence, as was seen during the state's recent drought from 2011 to 2017, which resulted in some of the most severe depletion of Central Valley aquifers and inelastic land subsidence. Future climatic conditions in the watersheds surrounding the Central Valley are expected to affect hydrology and water resources management. For example, the volume of water stored in the Sierra Nevada snowpack (one of the state's key "reservoirs") each spring has declined by about 10 percent in the last century and is expected to decline further. As a result, the state passed the Sustainable Groundwater Management Act (SGMA) in 2014 and is developing planning tools to cope with these increased demands and growing uncertainty in supplies.

The California Department of Water Resources (DWR) has developed the California Central Valley Groundwater-Surface Water Simulation Model (C2VSim) as a tool to aid in water management planning. This complex finite element model covers the entire Central Valley of California and is being used to aid in water management planning. The Integrated Water Flow Model (IWFM) couples a 3-D groundwater simulation process with 1-D land surface, river, lake, unsaturated zone, and small-stream watershed processes. It also dynamically calculates crop water demands; allocates contributions from precipitation, soil moisture, and surface water diversions; and calculates the amount of groundwater pumping required to meet the remaining demand. The model is being used for integrated regional water management plans, planning studies, groundwater storage investigations, assessing infrastructure improvements, evaluating ecosystem enhancement scenarios, conducting climate change studies, and assessing the impacts of changes to water operations. C2VSim is an integrated numerical model that simulates water movement through the linked land surface, groundwater, and surface water flow systems in California's Central Valley. Figure 7-12 shows the extent of the C2VSim model. Model input was compiled from DWR's extensive Central Valley land and water use data, which have been continuously collected since 1921. Figure 7-13 shows the C2VSim model grid.

The model uses a detailed database of monthly precipitation, land use, crop acreage, river inflow, and surface water diversion information from October 1921 through September 2009 to calculate historical water use, groundwater pumping, and changes in aquifer storage, and can be extended to simulate the response to projected future stresses. The C2VSim model output can be summarized to produce water budgets for each of 21 model subregions, five hydrologic regions, or the entire model area. Figure 7-14 is an example of input to the C2VSim model, showing the distribution of agricultural pumping in C2VSim.

Model results show that annual agricultural water demands increased from less than 6 million ac-ft (MAF) in the 1920s to more than 14 MAF by the 1960s, remaining at that level through the 2000s, and annual urban water demands increased more than tenfold from less than 0.2 MAF in the 1920s to more than 2 MAF in the 2000s. The surface water delivery volume is greater than the groundwater pumping volume in all but the driest years. Surface water deliveries peaked at more than 16 MAF in 1980 and were greater than 14 MAF/yr for 17 of the 40 years from 1970 to 2009. Groundwater pumping rose in years of surface water shortages, generally fluctuating between approximately 8 and 14 MAF/yr, peaking at 17 MAF in the 1977 drought. Between 1921 and 2009, groundwater withdrawal exceeded replenishment by approximately 200 MAF, causing the water table to drop as much as 200 ft in some parts of the Central Valley, and causing land-surface subsidence in several areas (over 30 ft in some areas). Lowering of the water table and changes in river flow rates have also reduced groundwater discharges to rivers, reducing summer flows with resulting increases in water temperatures and declines in water quality.

The C2VSim model will be useful for addressing several key water management questions in the Central Valley as the state progresses to comply with SGMA requirements

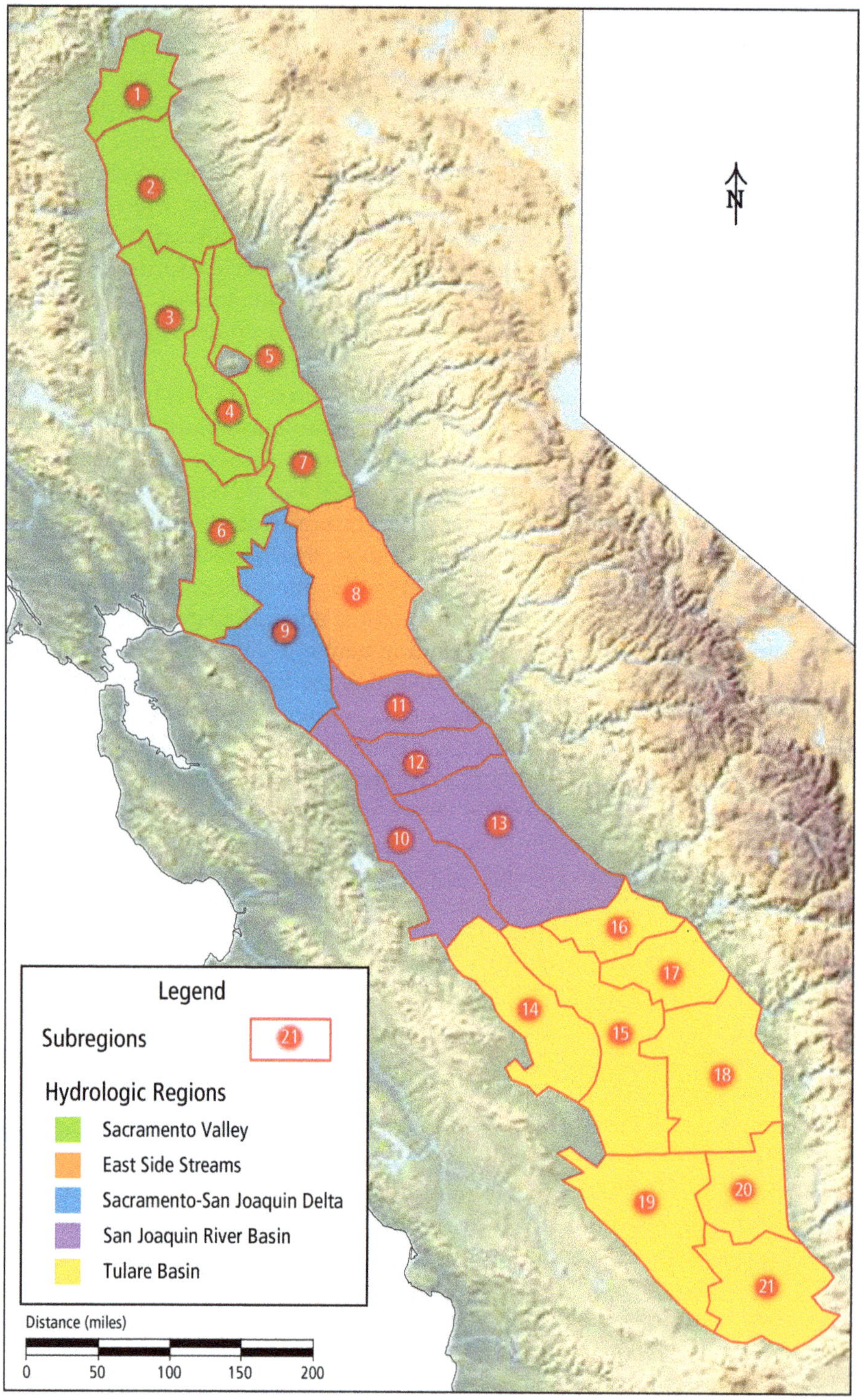

Source: Brush and Dogrul 2013

Figure 7-12 California Central Valley Groundwater—Surface Water Simulation Model (C2VSim) model area

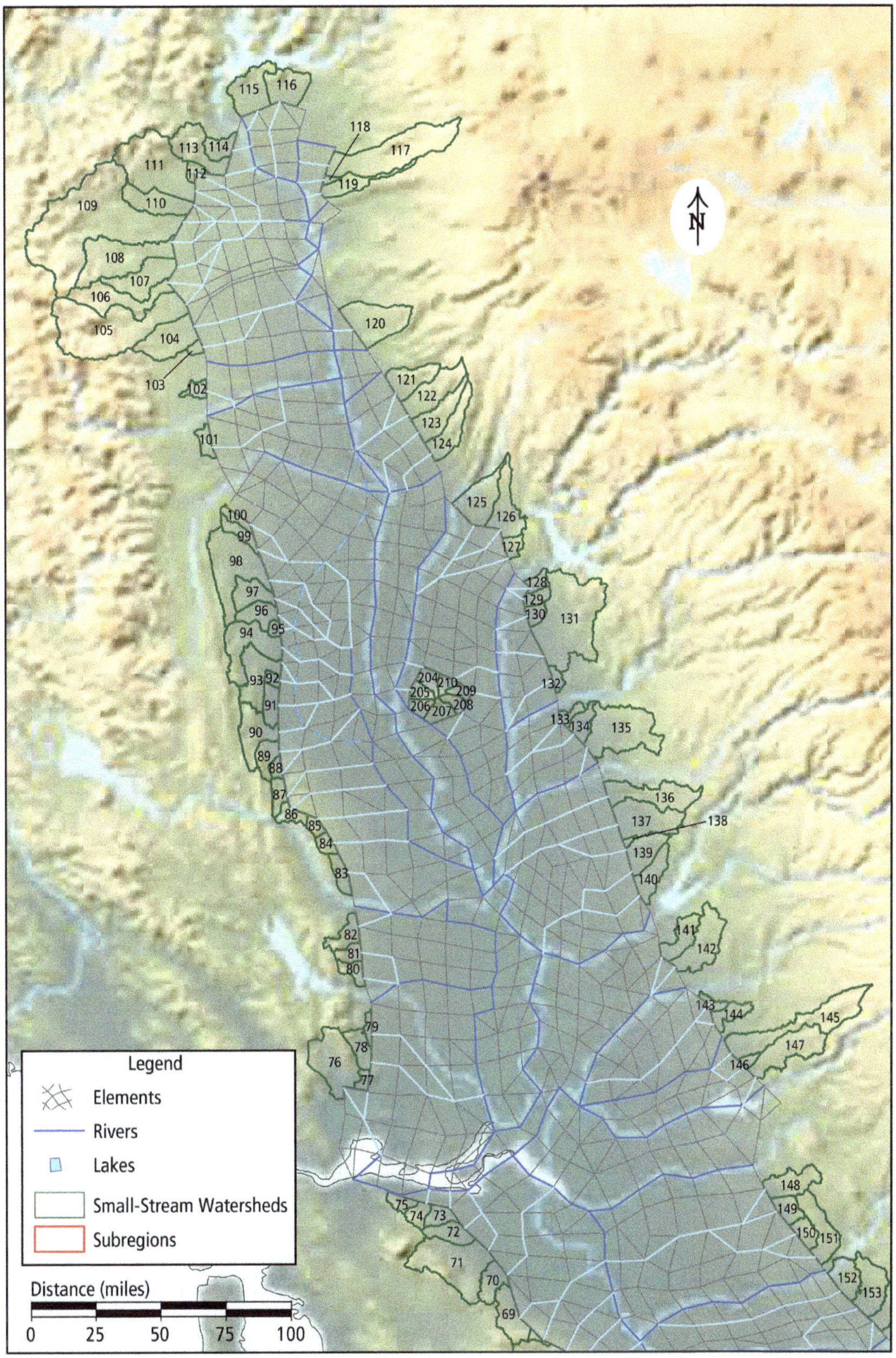

Source: Brush and Dogrul 2013

Figure 7-13 California Central Valley Groundwater—Surface Water Simulation Model (C2VSim) model grid

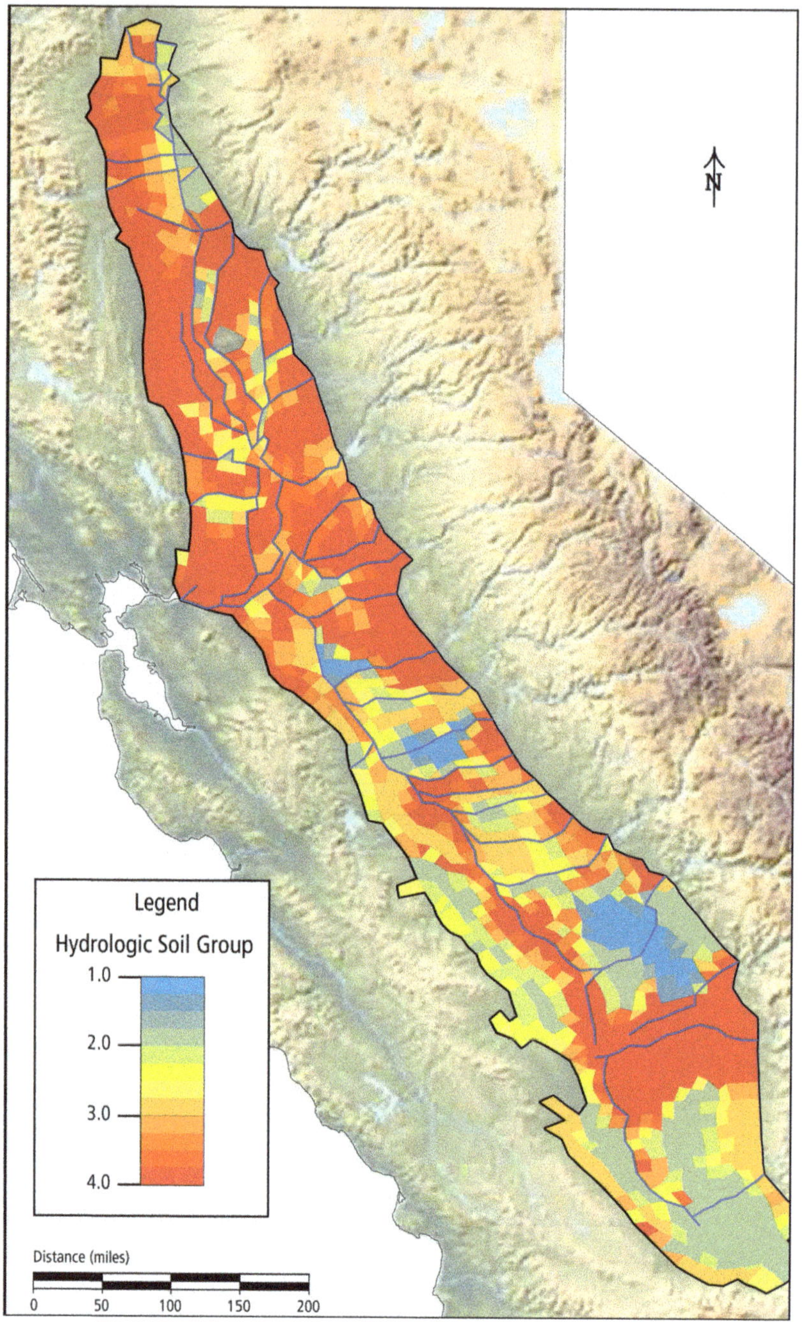

Source: Brush and Dogrul 2013

Figure 7-14 California Central Valley Groundwater—Surface Water Simulation Model (C2VSim) example model input—agricultural pumping distribution

for sustainable groundwater basin management. Figure 7-15 is an example of output from C2VSim. C2VSim can be used to estimate groundwater pumping rates, which are not measured or reported in the Central Valley. The model can also be used to understand water flows between rivers and groundwater aquifers, which are essential for evaluating the impacts of many conjunctive use and water transfer programs. The C2VSim–CG model is currently being used as the basis for the groundwater flow component of CalSim 3. It has also been used to investigate how Sacramento Valley water transfers may affect the San Joaquin Delta flows and how an extended drought may impact groundwater levels.

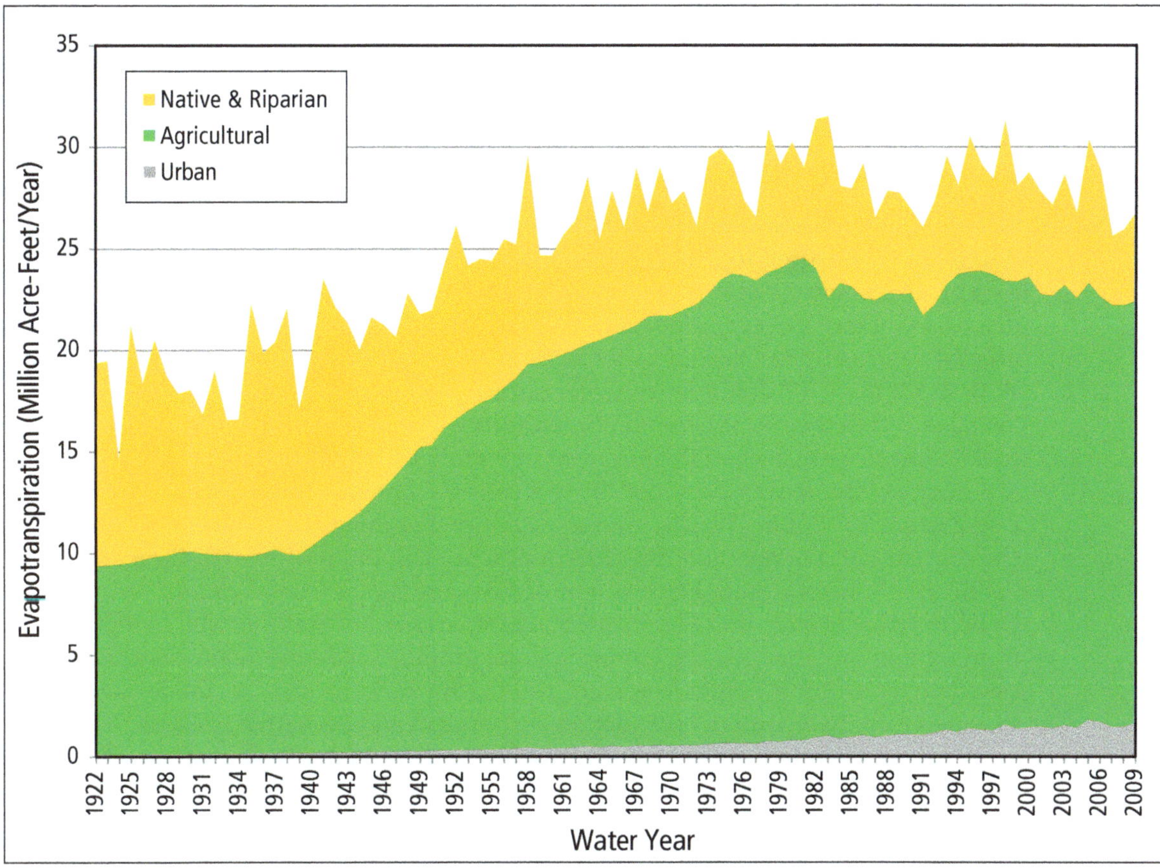

Source: Brush and Dogrul 2013

Figure 7-15 California Central Valley Groundwater—Surface Water Simulation Model (C2VSim) example model output

Both C2VSim versions will also be useful tools for integrated regional water management plans, planning studies, groundwater storage investigations, assessing infrastructure improvements, evaluating ecosystem enhancement scenarios, conducting climate change studies, and assessing the impacts of changes to water operations.

Project Name: California Central Valley Groundwater—Surface Water Simulation (C2VSim) Model

Agency: California Department of Water Resources (DWR)

Location: Central Valley, Calif.

Project objectives: Sustainable management, drought planning, wellhead protection, land use

Dimensions: Fully 3-D

Temporal coverage: Transient, variable time steps, multiple years

Computations: Finite element

Processes simulated: Groundwater flow, integrated groundwater–surface water flow

Software: Integrated Water Flow Model (IWFM) v.3.02

Case Study 7: Suffolk County Department of Health Groundwater Models

The aquifers beneath Suffolk County are sole source, providing drinking water for more than 1.5 million residents (see Figure 7-16). Since the mid-1990s, Suffolk County has used 3-D numerical groundwater flow models to help manage the aquifer system (see Figure 7-17). The models, which were developed to support source water assessments, have been adapted over time to address dozens of objectives, including sea level rise, saltwater intrusion, contaminant transport and containment, land use and zoning planning, and developing measures to reduce groundwater discharge of nitrogen to surface waters. As approximately 74% of the residents in the county are served by on-site wastewater disposal systems (OSSDS), elevated nitrogen concentrations are of concern to both potable groundwater supplies as well as the estuaries and embayments surrounding the county. With more than 700 community water supply wells and hundreds of monitoring wells installed across the county, frequent monitoring and protection of the aquifer to promote sustainable use has been at the forefront in Suffolk County for decades.

Critical input was provided by a steering committee made up of Suffolk County Department of Public Works, Suffolk County Planning Department, Suffolk County Department of Energy and the Environment, US Environmental Protection Agency (USEPA), New York State Department of Environmental Conservation, New York State Department of Health, and New York State Department of State, as well as representatives from local towns and villages, environmental groups, academic institutions, and building associations. County employees were trained on how to use the model and regularly execute groundwater model simulations to perform an alternative analysis.

Suffolk County has developed a Comprehensive Water Resource Management Plan that protects the county's groundwater and surface water resources and public drinking water supplies. The report is a publicly available document. Groundwater modeling informed the conclusions presented in the plan, including:

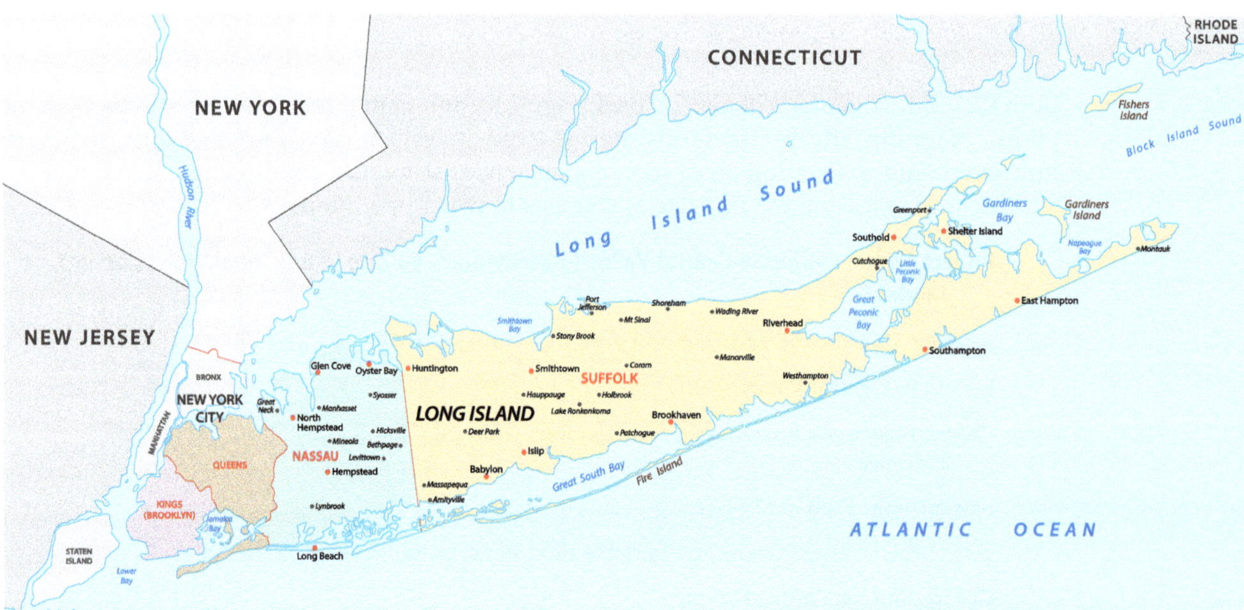

Source: Rainer Lesniewski / Shutterstock.com

Figure 7-16 Suffolk County, N.Y.

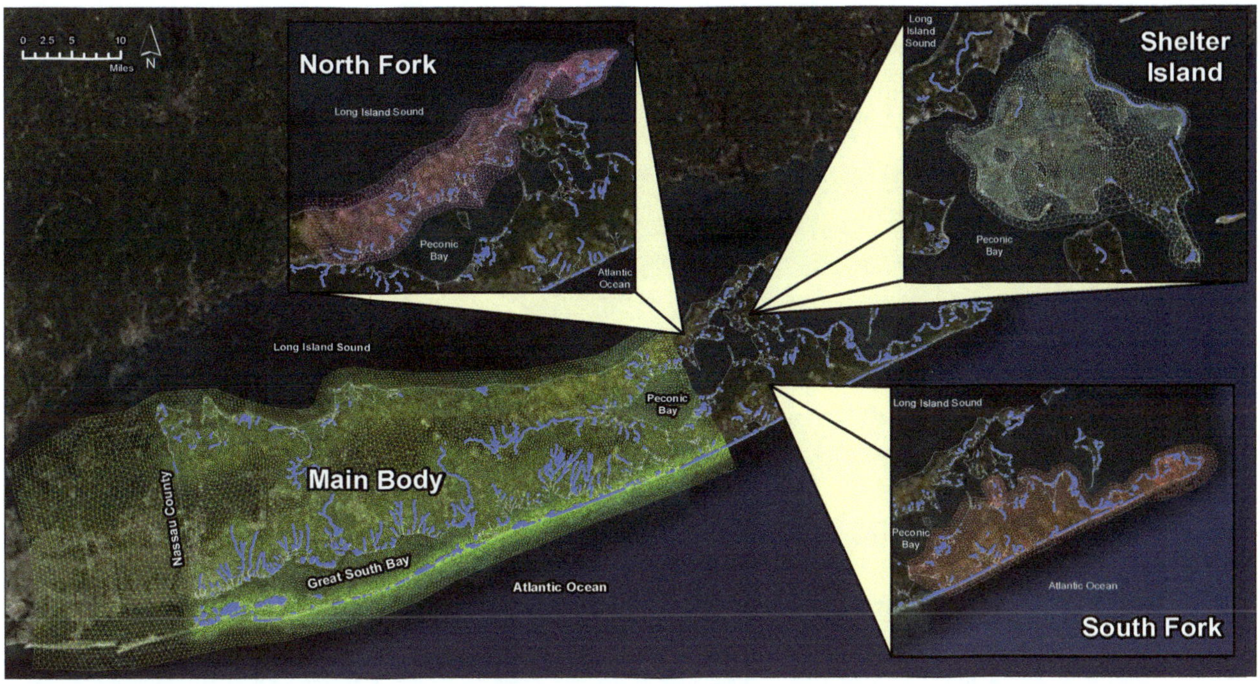

Source: Daniel O'Rourke, CDM Smith

Figure 7-17 Suffolk County groundwater model domains

- development of updated source water assessments for more than 500 public supply wells;
- estimation of sustainable pumping rates with respect to sea level rise, saltwater intrusion, and upconing;
- delineation of contributing areas of groundwater baseflow/underflow to the county's coastline and all major streams and rivers; and
- study of the relationship between land use, density, wastewater management, and nitrogen levels in Suffolk County.

The groundwater model was used to delineate the 50-year recharge areas to each of the county's community supply wells as well as the 25-year contributing areas to streams and embayments (see Figure 7-18). This work was used to help plan future development, quantify risk to natural resources, and update source water assessments and wellhead protection plans. Simulations were run to assess nitrogen movement through the aquifer and discharging to surface water bodies. These simulations were used to define areas of significance and surgically target solutions to reduce nitrogen concentrations and improve water quality.

Additionally, an assessment of the impacts of unsewered areas of various housing densities and OSSDS systems on nitrogen concentrations in groundwater was performed for a sparsely developed area within the county using the county's groundwater model and estimated effluent nitrogen concentrations for a range of treatment alternatives (from standard septic to membrane bioreactor (MBR)/Nitrex). Model simulations, which simulated the application and transport of nitrogen mass associated with each OSSDS, were run both with and without additional nitrogen loading associated with fertilizer application. Simulation results indicated that housing densities of ¼ to ½ ac yield nitrogen concentrations that exceed 6 mg/L when using standard septic systems, while 1-ac density

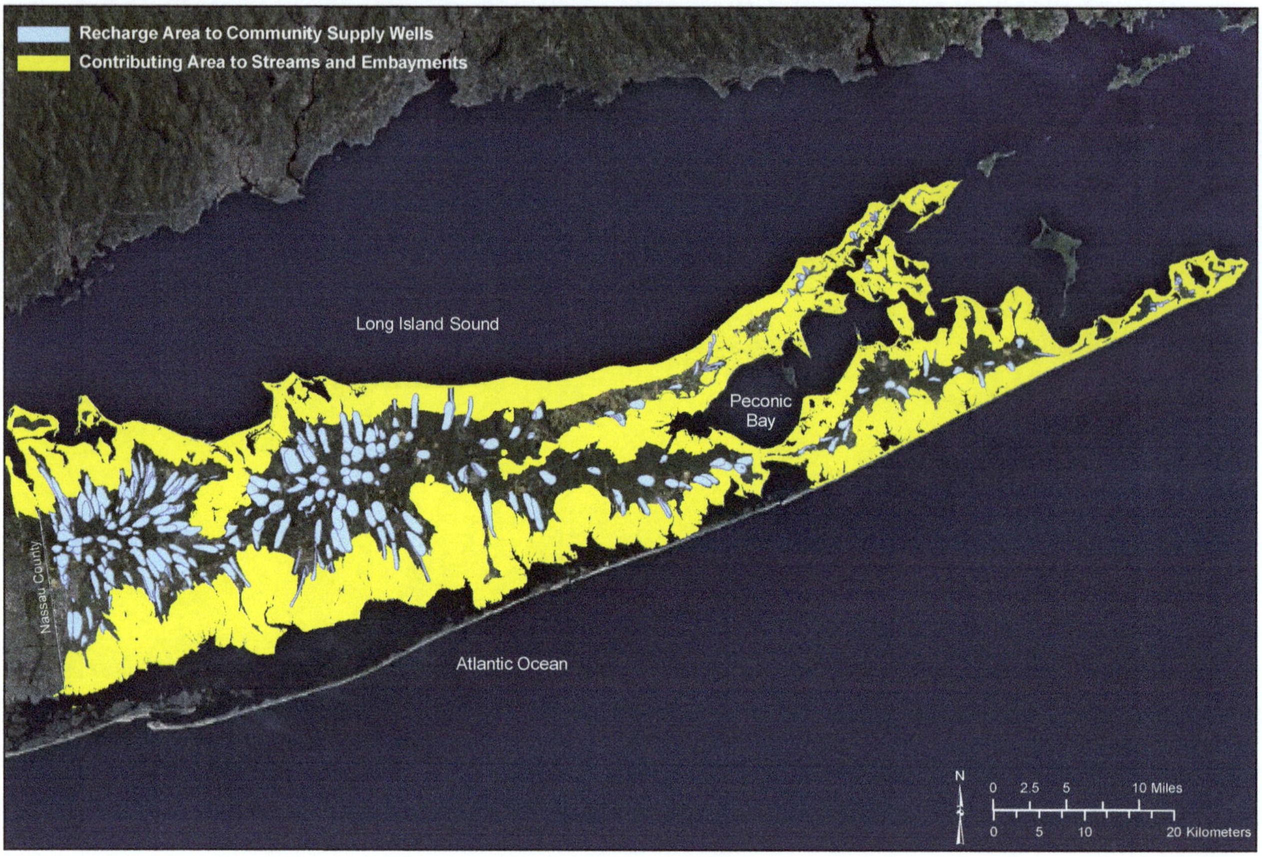

Source: Suffolk County Comprehensive Water Resources Management Plan (Suffolk County Department of Health Services, 2015)

Figure 7-18 Model-simulated contribution areas to wells and water bodies

resulted in concentrations below the target value. Utilization of alternative treatment systems yielded lower concentrations. The results will be used by the county to assess potential regulation changes to minimum lot size requirements for future development in unsewered areas.

The groundwater models were used to predict the impacts of a 3-ft rise of sea level (by 2100) on the groundwater table (see Figure 7-19). There are thousands of residential parcels located along the coastline of Suffolk County that are served by OSSDS. Sea level rise can have a direct impact on the treatment capacity of those systems as the separation distance from the disposal system and the water table will decrease in coastal communities as sea level increases. Groundwater model simulations were used to identify the areas where these impacts are expected to be most dramatic and to help county planners determine where future sanitary sewer projects will be most effective.

Model stewardship has been ongoing over the past 20 years through collaboration with all of the stakeholders. While initial model objectives were modest in scope, continued reliance on the model (and technological advances in computing power and data storage) have led to expanded capabilities.

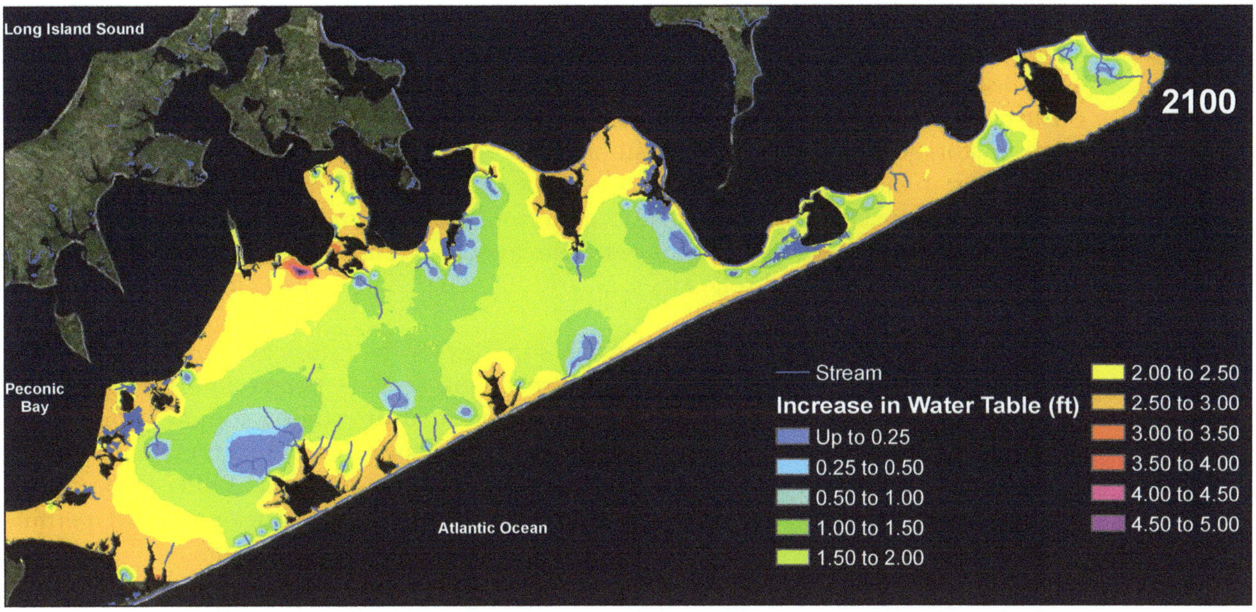

Source: Suffolk County Comprehensive Water Resources Management Plan (Suffolk County Department of Health Services, 2015)

Figure 7-19 Model-simulated increase in water table elevation because of sea level rise

Project Name: Suffolk County Comprehensive Water Management Plan

Agency: Suffolk County Department of Health

Location: Suffolk County, N.Y. (Long Island)

Project objectives: Sustainable management, drought planning, wellhead protection, nitrogen loading assessments, zoning

Size of model domain: Four models, 1.2 million ac

Dimensions: Fully 3-D

Temporal coverage: Transient, variable time steps, multiple years

Computations: Finite element

Processes simulated: Groundwater flow, contaminant transport, saltwater movement / dual density, sea level rise, discharge to streams (baseflow)

Number of drinking water wells included: >700

Software: DYNSYSTEM

Case Study 8: Hollywood, Fla., Salinity Barrier

This 3-D, dual-density model simulates the injection of treated wastewater to provide a barrier to inland movement of the saltwater interface toward public supply wells. The concept was to inject reclaimed water to retard saltwater intrusion created by the lowering of water levels by canals. Because the concept was reclaimed wastewater of the quality used for irrigation, regulatory issues were raised about the potential for risk to be introduced into the aquifer. The inclusion of both risk and contaminant transport (represented as rotavirus concentrations in injected reclaimed water) makes this a unique and complex application to water supply development and sustainability.

The City of Hollywood, Fla., is in southeast Florida, just south of Fort Lauderdale. Hollywood's primary groundwater resource is the surficial aquifer system, which includes the Biscayne aquifer and provides most of the freshwater for the city's population usage. The city's potable water treatment plant has a capacity of 35.7 mgd, with 11 million gal of on-site storage and two 1 million gal storage tanks in the city. At the time of the investigation, the water treatment plant had a consumptive use permit limiting Biscayne aquifer withdrawals to 20.67 mgd to be withdrawn from the Biscayne aquifer. Water is supplied to the city's 148,000 residents, plus residents in adjacent areas of Dania Beach, Miramar, Pembroke Park, and Pembroke Pines. Nearly 200,000 people and one international airport are currently served.

There are two issues regarding water supplies for the future growth in the city. The first is the quantity of water available for treatment. Because people in urbanized South Florida use groundwater supplies that are replenished directly by summer rainfall, the quantity of water available is finite and the quality must be protected for the end users—the public and the ecosystem. The second is that water supplies in South Florida are regulated by the South Florida Water Management District. The District issues water use permits based upon availability of the resource. These withdrawals limit both annual average and maximum daily withdrawals from the aquifer by the utility. Periodic renewal of the water use permits allows the water systems to adjust the quantities for withdrawal based on growth and/or prior experience. As a result of restrictions imposed by the District under its water resource availability rule approved in 2007, the city cannot capture more Biscayne aquifer water despite expected increases in demand unless it comes from a source such as direct rainfall, reclaimed water, or desalination. In addition, the city has battled issues with saltwater intrusion for years. As a result, the city looked at an option that would allow it to recapture water supplies using reclaimed water injected into the Biscayne aquifer, east of the current wells, to slow saltwater intrusion while stacking water in the city's well fields. The injection of treated water or wastewater into a Class I sole source drinking water aquifer generated concerns about the migration of water of "impaired quality" to potable water supply wells (all public water supplies are treated with lime softening, filtration, disinfection, and possibly, ion exchange, or nanofiltration). Regulatory barriers resulting from the perceived risks of adverse health effects from pathogens such as viruses have limited the concept of impaired water resources from being used for groundwater replenishment programs.

The limestones and surficial sands are recharged by rainfall that falls primarily from June to October each year and migration from the Everglades (as described by Merritt 1996; Virogroup 1994; Fish and Stewart 1990; Causaras 1985). Generally, the water level in the Biscayne aquifer averages 1–2 ft ngvd, except during extremely wet and dry periods (Merritt 1996); however, despite the low elevation, it remains one of the most productive aquifers in the world and supplies more than 6 million people with fresh, raw water supplies.

Three computer models were used to generate the Hollywood salinity barrier results: MODFLOW, MT3D, and SEAWAT. SEAWAT provides a convenient tool for the simulation of variable-density flow in that it utilizes all the major features of MODFLOW and MT3D. Except for minor changes in the MODFLOW Basic Package input file, input files for SEAWAT are identical to the standard MODFLOW/MT3D input files, and they can be generated using any existing preprocessor programs for MODFLOW/MT3D. The selection of SEAWAT for this project was mainly based on the convenience of data conversions since SEAWAT uses standard MODFLOW and MT3D input files (Hazen and Sawyer and CDM/Missimer International 2000). The objective of this study was to examine the risk assessment using computational modeling with MODFLOW and MT3D groundwater transport simulation.

The model consisted of five layers that represent the surficial aquifer, a semi-confining layer, the Shallow Production Zone of the Biscayne aquifer, the Deep Production Zone of the Biscayne aquifer, and a deeper semi-confining layer, respectively. These layers were based on the study of geologic cross-sections available in the area. During the flow calibration, the model utilizes a quasi–3-D approach with the confinement between the shallow and deep production zones being represented by a leakance factor. True layer thickness was added to the model after the flow calibration because solute transport simulations require true 3-D model structure. The primary geologic parameters are locations of model boundaries and thickness of geologic unit. Physical and chemical parameters such as chemical reaction, porosity, specific yield, storage, transmissivity, and hydraulic conductivity were also included. Within the model area, the opinion of Virogroup (1994), based on drilling data, was that the layers were relatively homogeneous. Note that the water table was at an elevation of 2 ft, which is the king tides.

Some initial conditions given for the MODFLOW model were the total withdrawal rate of 24.4 mgd, split between two well fields. The north well field withdrawals were limited to 15 mgd and the south well field to 9.4 mgd. The canals in the area are considered constant head boundaries with 2 ft as initial head condition. The Florida Turnpike has a canal alongside, thus allowing the western boundary to be treated as a constant head. The northern and southern boundaries of the model were represented by C-10 and C-11 canals, respectively, which are both considered constant head boundaries (see Figure 7-20). The Atlantic Ocean was used as an eastern constant head boundary. All boundaries were considered to have constant piezometric head of 2 ft, which is the high tide condition. This value may be altered throughout the year, but there are no actual data on the tidal levels at the upper reaches of these canals. Since groundwater moves slowly, the high tide condition was used as this appears to match findings by others (Bloetscher et al. 2019, 2014a).

Initial values for hydraulic parameters were derived from the regional model, including transmissivity, leakance factor, and storage coefficients. Detailed description of these parameters can be found in a report by Virogroup (1994). These parameters needed to be fine-tuned in the vicinity of the well IW-1 using the data collected from the injection test. The model was initially calibrated for flow only; the solute transport was added after the flow was calibrated and then both water quality and water buildup were compared to measured values. Although these values in most of the model area were kept unchanged,

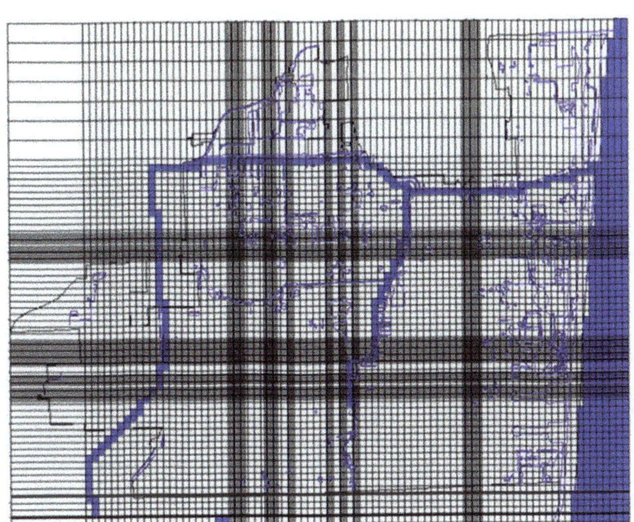

Source: Phonpornwithoon 2014

Figure 7-20 Initial grid boundary in the Biscayne aquifer

their values in the vicinity of the injection test site were adjusted as the results of model calibration in this study.

A constant initial chloride concentration of 200 mg/L was used for model layers 1 and 2, the surficial aquifer, and the semi-confining unit. The initial chloride concentration of model layer 4 was derived from data collected at the site (Hollywood Coastal Salinity Barrier Pilot Project, Second Phase Drilling Interim Technical Memorandum, Hazen and Sawyer/Missimer International 2000). The initial chloride concentration of model layer 3 was derived based on logarithmic weighting between the values of layers 2 and 4. The initial concentration in model layer 5 was assumed to be the same as that in layer 4. The source and sink term in the flow model in the transport equation represents solute mass dissolved in water that enters and exits the flow domain through fluid sources and leaves through fluid sinks (Zheng and Bennett 1995). SEAWAT was used to simulate aquifer response under aquifer injection test conditions. The model was calibrated for the Shallow and Deep Production Zones of the Biscayne aquifer by varying the aquifer's hydraulic parameters such as transmissivity and leakance until the simulated changes of water levels matched the changes of water levels during the preceding 12 months. Flow and solute transport models of the area were developed using tested and verified codes capable of simulating aquifer response to groundwater withdrawals/injections and density-dependent groundwater flows. The flow model was used for initial model calibrations because the lower simulation time allowed more rapid processing of the iterative runs required to determine an optimal solution. The aquifer hydraulic parameters resulting in the best fit between simulated and observed aquifer response were determined through the calibration process. The solute transport model was subsequently developed to incorporate chemical mass transfer and density-dependent flow components using the hydraulic parameters determined from the calibration model.

Once the model was calibrated, the contaminants were added. The regulatory concern was pathogens. Based on a review of the available literature, a common cause of serious gastroenteritis in humans that is easily transmitted is rotavirus (Gerba et al. 1996; Maldonado and Yolken 1990). The concentration of rotavirus in water samples was calculated assuming the City of Hollywood treatment level, which is secondary treatment with filtration and a high level of disinfection. Based on the only study found, the average

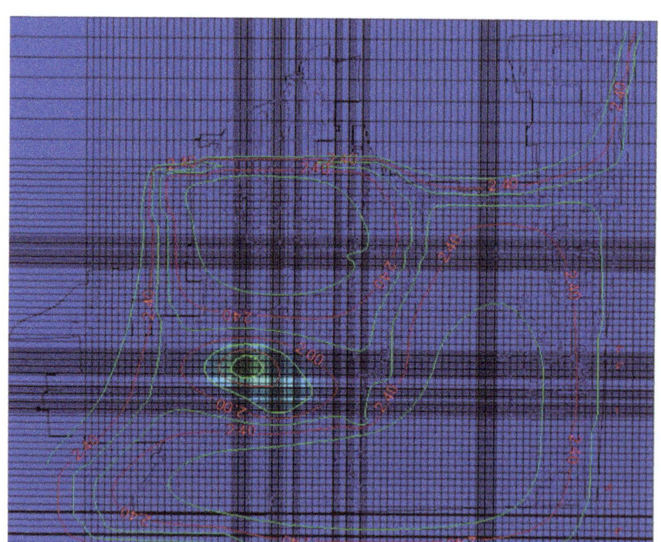

Source: Phonpornwithoon 2014

Figure 7-21 Groundwater table drawdown, September 2011 (wet season)

concentration of rotavirus in raw wastewater was 16.6 pathogens (PFU) per liter with a detection rate of 33% (Li et al. 2011). The worst-case scenario was 300 PFU/L and the detection rate was 100%. The reduction in concentration was 2.08 logs with secondary treatment (Li et al. 2011). Virus removal in the secondary treatment was mostly attributed to the adsorption of microorganisms on the activated sludge per Gerba et al. (1996). Filtration achieved another 0.72 log removal (Li et al. 2011). Chlorine can add another percentage of removal, but to be conservative, in the absence of a stated number, it was assumed that (combined) a total of 2.8 logs of removal was present. The injection well used for this transport modeling was estimated to deliver 1 mgd into the recharge zone, which is 200 ft below ground level. Based on a flow of 1 mgd, a value of 1.6 million PFU were injected assuming the worst-case scenario on a daily basis:

$$\text{Calculation:} \frac{3.28 \times \dfrac{10^6\,\text{L}}{\text{mgd}} \times 300\,\text{PFU/L}}{10^{2.8}} = 1.6\,\text{million}\,\text{PFU/d} \qquad \text{(Eq 7-1)}$$

For ease in modeling, it was assumed that the daily amount of rotaviruses delivered was 2 million PFU/mgd of injection water.

It was assumed that the injection wells will be operated continuously for many years, thereby allowing the program to reach equilibrium by $\overline{C}(x,y,z,t = \infty)$. However, the groundwater program cannot assume that the time element was infinity. The model was re-run for a two-year period, with recharge via rainfall. Because of its high shedding incidence, high exposure, and low infective dose, rotavirus was modeled, and an appropriate decay rate was incorporated to use the concentration contours and convert them to a risk contour. Using the dose response curve, the resulting concentrations from the groundwater model were directly translated to risk contours. These contours decrease with increasing distance from the injection point, although because of the sharp increase in the dose-response curve developed by the Predictive Bayesian equation, the risks between 0 and 0.1 occur over a very short horizon. The data from MT3D simulation were exported to GIS as a shape file, which also included concentration level per contour. Using GIS, each contour color was converted to risk contours based on predicted concentration. It should be noted that these are log concentration contours. After the data was exported onto GIS layers, the distance between each contour layer and injection well was measured using the measuring tool in GIS (see Figure 7-21).

Having completed one analysis, different scenarios of virus loading rate for rotavirus were simulated along with enterovirus to compare the risk of injection associated with each scenario. See Table 7-2 for the different scenarios and their respective decay rate and distance of risk contour from the injection well. The two worst-case scenarios were simulated with one assumed rotavirus decay rate at zero and another at 0.1027 per day. The worst-case scenario used the detection rate of 100% and 300 PFU/L and no decay. The third scenario is the average case, which used the detection rate of 33% and 16 PFU/L with decay rate of 0.1027 per day. The fourth scenario uses enterovirus with viral loading rate of 1 PFU/L, which is equal to concentration of approximately 6,000 per 1 million gal of injection water per day (Bitton et al. 1983).

Figure 7-22 shows the results of the risk contours for different scenarios and their respective distances from the injection well based on Table 7-2. The City of Hollywood's production well fields are located approximately 3 mi from the injection well. Assuming worst-case scenario number 2, the risk contour is estimated to be about 2.3 mi from the production well fields, which might suggest that the injection program as feasible. Because

Table 7-2 Results from different scenarios of viral loading rate

Scenario	Viral Loading Rate	Decay (day^{-1})	Distance from Injection Well at 10^{-6} (mile)	Half-Life
Worst case – no decay	2 million	–	2	–
Worst case	2 million	0.1027	0.7	6.75
Average case	30,000	0.1027	0.3	6.75
Nonspecific enterovirus	6,000	0.0456	0.25	15.2
Virus removed by reverse osmosis treatment	0	0.1027	0	6.75

Source: Rotavirus decay from Yates et al 1985. Enterovirus decay from Bitton et al. 1983
Note: Dashes indicate no data.

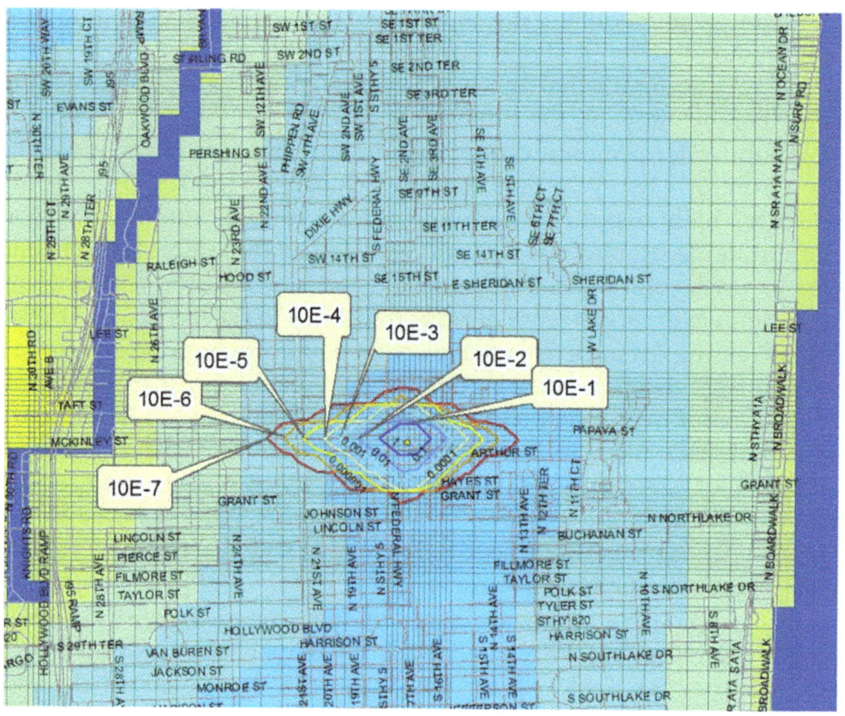

Source: Phonpornwithoon 2014

Figure 7-22 Risk contours related to rotavirus risk in raw water supply wells given injection of reclaimed wastewater

of lower viral loading rate for enterovirus, the risk contour from contaminant transport modeling shows the migration of virus for 0.25 mi from the injection well (see Figure 7-23). The results indicated that using reclaimed water as a salinity barrier does not pose a public health risk in this scenario, which was a goal of the study.

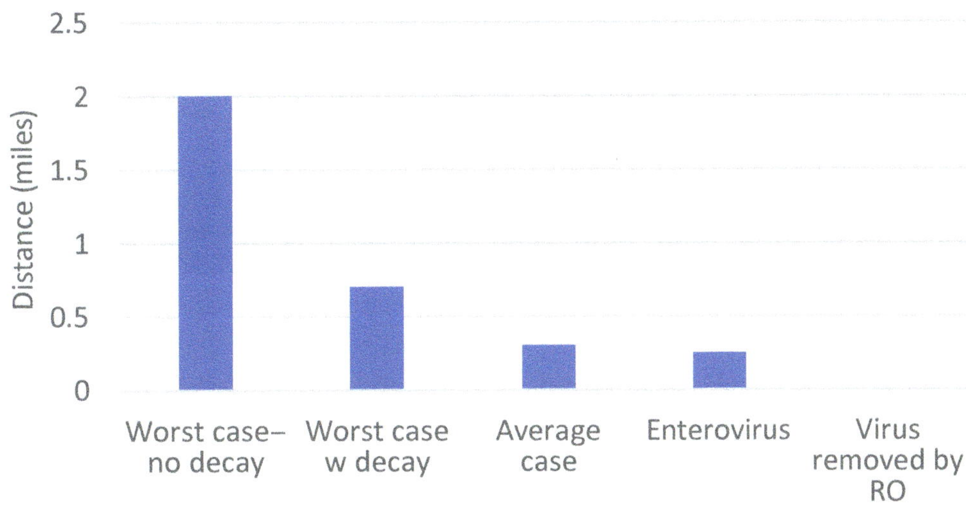

Source: Phonpornwithoon 2014

Figure 7-23 Comparison of scenario vs. distance of risk contour from injection well

Project name: City of Hollywood Salinity Barrier

Agency: City of Hollywood, Fla.

Location: Hollywood, Fla.

Project objectives: Determine risk potential with recharge of Biscayne aquifer with reclaimed wastewater

Size of model domain: 30 mi^2

Dimensions: Fully 3-D

Temporal coverage: Transient, monthly time steps, two years

Computations: Finite difference

Processes simulated: Groundwater flow, contaminant transport, saltwater movement / dual density, sea level rise, discharge to/from canals (baseflow)

Number of drinking water wells included: 24

Software: Groundwater Vistas processor

REFERENCES

Bitton, G.; Farrah, S.R.; Ruskin, R.H.; Butner, J.; and Chou, Y.J. 1983. Survival of Pathogenic and Indicator Organisms in Ground Water. *Groundwater*. 21:4:405–410.

Bloetscher, F.; Meeroff, D.E.; Long, S.C.; and Plummer, J.D. 2020. Demonstrating the Benefits of Predictive Bayesian Dose-Response Relationships Using Six Exposure Studies of *Cryptosporidium parvum*. *Risk Analysis*. 40:11:2422–2461.

Bloetscher, F.; Meeroff, D.E.; and Phonpornwithoon, P. 2019. Assessing Risk of Injection of Reclaimed Water into the Biscayne Aquifer for Aquifer Recharge Purposes. *Journal of Geoscience and Environment Protection*. 7:7:184–201.

Brush, C.F. and Dogrul, E. 2013. *User's Manual for the California Central Valley Groundwater-Surface Water Simulation Model (C2VSim), Version 3.02-CG*. Sacramento, Calif.: California Deptartment of Water Resources.

Causaras, C. 1985. Geology of the Surficial Aquifer System, Broward County, Florida. Reston, Va.: USGS (US Geological Survey).

Crandall, C.A.; Kauffman, L.J.; Katz, B.G.; Metz, P.A.; McBride, W.S.; and Berndt, M.P. 2009. *Simulations of Groundwater Flow and Particle Tracking Analysis in the Area Contributing Recharge to a Public Water Supply Well near Tampa, Fla. 2002–2005*. Tallahassee, Fla.: USGS.

Fish, J.; and Stewart, M. 1990. *Hydrogeology of the Surficial Aquifer System, Dade County, Florida*. Tallahassee, Fla.: USGS.

Gerba, C.P.; Rose, J.B.; Haas, C.N.; and Crabtree, K.D. 1996. Waterborne Rotavirus: A Risk Assessment. *Water Research*. 30:12:2929–2940.

Hazen and Sawyer and CDM/Missimer International 2000. *Hollywood Coastal Salinity Barrier Pilot Project Phase I Potable Water Injection Testing Technical Memorandum*. Boca Raton, Fla.: Hazen and Sawyer.

Honeyman, R. 2014. *Modeling of the Page Aquifer: An Assessment of Groundwater Availability*. Bismarck, N.D.: North Dakota State Water Commission.

Jagucki, M.L.; Katz, B.G.; Crandall, C.A.; and Eberts, S.M. 2009. *Assessing the Vulnerability of Public-Supply Wells to Contamination: Floridan Aquifer System Near Tampa, Florida*. Reston, Va.: USGS.

Langel, R. and Gannon, J.M. 2011. *Use of Numerical Modeling to Evaluate the Quantity and Quality of Induced Surface Water Recharge for the City of Dunlap*. Des Moines, Iowa: Iowa Department of Natural Resources.

Li, D.; Gu, A.Z.; Zeng, S.Y.; Yang, W.; He, M.; and Shi, H.C. 2011. Monitoring and Evaluation of Infectious Rotavirus in Various Wastewater Effluents and Receiving Waters Revealed Correlation and Seasonal Pattern of Occurrences. *Journal of Applied Microbiology*. 110:5:1129–1137.

Maldonado, Y.A. and Yolken, R.H. 1990. Rotavirus. *Bailliere's Clinical Gastroenterology*. 4:3:609–625.

Merritt, M.L. 1996. *Assessment of Saltwater Intrusion in Southern Coastal Broward County, Florida*. Tallahassee, Fla.: USGS.

Omaha Metropolitan Utilities District. 2012. *2011 Nebraska Ordnance Plant Groundwater Report*. Omaha, Neb.: Metropolitan Utilities District.

Omaha Metropolitan Utilities District. 2005. Phase II Platte West Well Field/Groundwater Modeling Study. Omaha, Neb.: Omaha Metropolitan Utilities District.

Phonpornwithoon, P. 2014. *Risk of Injection Using Reclaimed Water for Aquifer Recharge Using Rotavirus as Surrogate Contaminant*. Ann Arbor, Mich.: ProQuest Publishing.

Shaver, R.B. 1994. *An Analysis and Conceptual Hydrogeologic Model of a Till Aquitard Overlying the Spiritwood Aquifer in Southeastern North Dakota*. Bismarck, N.D.: North Dakota State Water Commission.

Shaver, R.B. and Pusc, S.W. 1992. Hydraulic Barriers in Pleistocene Buried-Valley Aquifers. *Groundwater*. 30:1:21–28.

Suffolk County Department of Health Services. 2015. *Suffolk County Comprehensive Water Resources Management Plan*. Hauppauge, N.Y.: Suffolk County Department of Health Services.

Taylor, M. and O'Rourke, D. 2009. *Overview of Lloyd Aquifer Representation in Nassau and Suffolk County Groundwater Models*. New York: CDM.

Tepper, D.H.; Morrissey, D.J.; Johnson, C.D.; and Maloney, T.J. 1990. *Hydrogeology, Water Quality, and Effects of Increased Municipal Pumpage of the Saco River Valley Glacial Aquifer: Bartlett, New Hampshire to Fryeburg, Maine*. Reston, Va.: USGS.

Virogroup. 1994. *Determination of Safe Yield from the Biscayne Aquifer Wellfield at the City of Hollywood Using Hydraulic and Solute Transport Models*. Cape Coral, Fla.: Virogroup.

Winner, M.D., Jr. and Coble, R.W. 1996. *Hydrogeologic Framework of the North Carolina Coastal Plain, Regional Aquifer-System Analysis—Northern Atlantic Coastal Plain*. Raleigh, N.C.: USGS.

Yates, M.V.; Gerba, C.P.; and Kelly, L.M. 1985. Virus Persistance in Groundwater. *Applied and Environmental Microbiology*. 49:4:778–781.

Zheng, C. and Bennett, G.D. 1995. *Applied Contaminant Transport Modeling*. New York: Van Nostrand Reinhold.

This page intentionally blank.

Chapter 8

The Future of Groundwater Management

This manual addresses tools and models that support improved and proactive management of groundwater resources. Background information is included about the characterization of aquifers as intended for use by a water utility, including those parameters of the aquifer needed to create a useful management plan and model of the aquifer system. Tools in use by many utilities include data needs for developing a groundwater budget, methods for delineating the aquifer, and fundamental concepts controlling well and aquifer yield. As states and municipalities have experienced significant decreases in groundwater levels and wellfield yields, groundwater and aquifer management is undergoing substantial changes with the utilization of regional management tools and groundwater models. In-depth groundwater budgeting, water balance and availability calculations, groundwater pumping, impacts to rivers and streams because of groundwater depletion, subsidence, and well field yields all support the move toward integrating utility water management. These tools will also help establish more targeted monitoring of changes in aquifers and well fields, and to move toward solutions that are relevant to each utility.

The authors of this manual hope that the content included in this manual, used conjunctively with the case studies provided in Chapter 7, can be used to answer the following questions: Is a sustainable water supply available for your community? How do we manage the groundwater to ensure it is sustainable? Modeling is a tool to help answer that question, but modeling is only as useful as those modeling and the inputs they use. As a result, models should be dynamic—they should be adjusted as conditions changes, more information is gained, or competition increases. They are only useful as long as they calibrate to the actual conditions experienced in the field. When they do not, adjustments are required, perhaps both in the model and in the pumping of the aquifer.

From a longer-view perspective, avoiding the initial scenario outlined in Chapter 1 for the Carolinas is one of the goals of this manual. With more recent publications by the US Geological Survey (USGS), the US Environmental Protection Agency (USEPA), and AWWA, the concern of managing groundwater for sustainability has become more critical. Many people still believe that groundwater is an endless supply, protected because it cannot evaporate, and easily recharged from rainfall. But this is not the case in most places. We can, and do, draw groundwater down faster than it can recharge.

The ANSI/AWWA G300 standard on source water protection exists not just for surface waters but to protect the quality of groundwater from contamination. Water quality is one piece of the issue. Addressing water quantity while negotiating the maze of water rights holders, downstream user needs, and competition for supplies are challenges that are expected to increase in the 21st century. States such as Texas, Arizona, Nevada, Utah, California, and Colorado have initiated planning efforts to accomplish this goal. Permitting has been initiated in some states as they suffer from diminished supplies to attempt to manage supplies for the future. Because utilities are perceived to have "deep pockets," they will be looked to for solutions.

Water security—the ability to have enough water to address the community's economic and social needs—is intrinsic in the holistic groundwater management strategy. Hence, modeling how much water that can sustainably be withdrawn will help with the supply. The examples in Chapter 7 were chosen to show increasing complexity, increasing use of the data, and potential risks to public health. The models have shown that water quality and risk can build upon a good groundwater management model. All the mathematics for contaminant transfer and risk are based on the accurate representation of groundwater movement.

Groundwater utilities are encouraged to develop their own models. They should be as complex as necessary. Tenets within the ANSI/AWWA J100 standard for utility management and federal regulations on water security should be included in the modeling effort as a means to identify critical venues. These steps will provide the ability for utilities to make better management decisions on water supply needs and treatment options required. Managing groundwater appropriately will help build sustainability into the system.

Today, many areas are overdrawing their aquifers without realizing it. That creates an economic issue that in time will require solutions. Potential solutions for decreasing groundwater supplies may be in the form of managed aquifer recharge or spreading basins that are increasingly used in regions with the appropriate land space, water accessibility and trades via "dry and wet" water wheeling and water market agreements, aquifer storage and recovery, and other technical and economically viable approaches. Modeling can help utility staffs to better manage their water supplies.

Emerging trends in groundwater management are similar to those in other arenas. There are significant amounts of data that can be obtained by federal, state, and local jurisdictions. There is so much information that the use of data analytics is needed to help identify key parameters for development of a regional groundwater model, and the ensuing management approaches for wells and well fields of an individual utility. As a result, the future of groundwater management can be summed as:

- understanding the issues in the context of continual changes in aquifers;
- understanding aquifer and groundwater system limits;
- defining sustainable groundwater withdrawals;
- monitoring to make changes if the assumption proves incorrect; and
- including groundwater supplies as part of a utility's sustainability planning.

REFERENCES

ANSI/AWWA. G300. Source Water Protection. Denver, Colo.: AWWA.

ANSI/AWWA. J100. Risk and Resilience Management of Water and Wastewater Systems. Denver, Colo.: AWWA.

This page intentionally blank.

Appendix A

Glossary of Concepts Associated with Aquifer and Wellfield Yield

Several data are needed to determine the well yield. These data are aquifer parameters. The most significant aquifer parameters are porosity, transmissivity, specific yield and specific retention, and hydraulic head and gradient. Porosity, specific yield, and specific retention describe the rock formation and quantities of water existing in the formation. Head and gradient determine how water moves through the formation and represent the mechanics of horizontal and vertical recharge to a well being pumped. Head and gradient are also used to analyze the transport of pollutants that may migrate to a well. Transmissivity indicates how easily water will move in the formation and is perhaps the most common term used by hydrogeologists.

Porosity. The ratio of openings (voids) to the total volume of a soil or rock is referred to as porosity. Porosity is expressed either as a decimal fraction or as a percentage:

$$n = \frac{V_t - V_s}{V_t} = \frac{V_V}{V_t} \qquad \text{(Eq A-1)}$$

Where:

n = porosity, as a decimal fraction
V_t = the total volume of a soil or rock sample
V_s = the volume of solids in the sample
V_v = the volume of openings (voids)

If the porosity determined using Eq A-1 is multiplied by 100, the result is porosity expressed as a percentage. Porosity values determine the maximum amount of water that a rock can hold when it is saturated. Only a part of the water in the pores is available to supply a well or a spring. Secondary porosity is important because it describes the extent to which the voids are interconnected, allowing water to move. Secondary porosity arises from the dissolution of cement between grains, the grains themselves, the rock matrix, etc.

Soils are among the most porous of natural materials because soil particles tend to form loose clumps and because of the presence of root holes and animal burrows. The porosity of unconsolidated sand and gravel depends on the range in grain size "sorting" and on the shape of the rock particles but not on their size. Fine-grained materials tend to be better sorted and have the highest porosity values. Limestone and sand both have high secondary porosity.

Specific yield and specific retention. Hydrologists divide groundwater into the portion that will drain under the influence of gravity, which is called specific yield, and the portion that is retained as a film on rock surfaces and in very small openings, which is called specific retention. The physical forces that control specific retention are the same forces controlling the thickness and moisture content of the capillary fringe. Thus:

$$n = S_y + S_r \qquad \text{(Eq A-2)}$$

$$S_y = \frac{V_d}{V_t} \text{ and } S_r = \frac{V_r}{V_t} \qquad \text{(Eq A-3)}$$

Where:

n = porosity
S_y = specific yield
S_r = specific retention
V_d = the volume of water that drains from a total volume of V_t
V_r = the volume of water retained in a total volume of V_t
V_t = total volume of a soil or rock sample

Hydraulic head and gradient. The depth to the water table affects the development of water supplies from unconfined aquifers. Where the water table is shallow, the land may become waterlogged during wet weather and unsuitable for most land uses (such

Table A-1 Selected values of porosity

Material	Primary Openings*	Secondary Openings*
Equal-size spheres (marbles)		
Loosest packing	48	—
Tightest packing	26	—
Soil	55	—
Clay	50	—
Sand	25	—
Gravel	20	—
Limestone	10	10
Sandstone (semi-consolidated)	10	1
Granite	—	0.1
Basalt (young)	10	1

Source: Heath 1983
* Values are given in percent by volume. Dashes indicate no data.

as houses). Where the water table is at great depth, the cost of constructing wells and pumping water for domestic needs may be prohibitively expensive.

The highest head in an aquifer occurs at the points of aquifer recharge. The position and the slope of the water table (or of the potentiometric surface of a confined aquifer) is determined by measuring the position of the water level in wells from a fixed measuring point. The position of the water table at each well must be determined relative to a datum plane that is common to all the wells. The datum plane most widely used is now NAVD88, an update from the 1929 datum. In either case, zero is commonly referred to as *sea level*.

Total head. The depth to water in a nonflowing well is subtracted from the elevation of the measuring point to determine the total head at the well. Total head, as defined in fluid mechanics, is the sum of elevation head, pressure head, and velocity head. Because groundwater moves relatively slowly, velocity head can be ignored. Therefore, the total head at an observation well involves only elevation head and pressure head. Pressure head exists only in confined aquifers and expresses the amount of artesian pressure the water is under. Groundwater moves in the direction of decreasing total head (i.e., downhill), which may or may not be in the direction of decreasing pressure head.

Hydraulic gradient. All other factors being constant, how fast groundwater moves depends on the hydraulic gradient. The hydraulic gradient is the change in head per unit of distance in a given direction. If the direction is not specified, it is in the direction in which the maximum rate of decrease in head occurs.

When the hydraulic gradient is expressed in consistent units (i.e., where both the numerator and the denominator are in meters or feet), any other consistent units of length can be substituted without changing the value of the gradient. The change in height over length (dh/dL) provides the aquifer or head slope.

Calculating hydraulic gradient and direction of groundwater movement. Both the direction of groundwater movement and the hydraulic gradient can be determined if data are available for three wells located in any triangular arrangement. These data are:

- the relative geographic position of the wells,
- the distance between the wells, and
- the total head at each well.

Hydraulic conductivity. Hydraulic conductivity replaces the term "field coefficient of permeability" and should be used to refer to the water-transmitting characteristic of material in quantitative terms.

The factors controlling groundwater movement were first expressed by Henry Darcy, a French engineer, in 1856. Unlike rivers and streams, groundwater tends to move relatively slowly, often measured in feet per day or year. Because of this slow movement, groundwater flow is said to be laminar; that is, water particles tend to follow discrete streamlines and not to mix with particles in adjacent streamlines. As a result, Darcy's Law says that the quantity of water, Q, is directly proportional to the head gradient, dh/dL. Thus, the units of hydraulic conductivity are those of velocity (or distance divided by time). Hydraulic conductivity not only varies by type of rock but may also be different from place to place in the same rock. If the hydraulic conductivity is essentially the same throughout an area, the aquifer is homogeneous. If the hydraulic conductivity differs from one part of the aquifer to another, the aquifer is heterogeneous.

Hydraulic conductivity may also vary by direction at any place in an aquifer. If the hydraulic conductivity is essentially the same in all directions, the aquifer is isotropic. If it varies by direction, such as differences between conductivity in the vertical and horizontal directions, the aquifer is anisotropic.

Although it is convenient in many mathematical analyses of groundwater flow to assume that aquifers are both homogeneous and isotropic, such aquifers are rare, if they exist at all. Hydraulic conductivity in most rocks and especially in unconsolidated deposits and in flat-lying consolidated sedimentary rocks is larger in the horizontal direction than in the vertical direction, usually by a factor of about 10.

Transmissivity. Transmissivity is the capacity of an aquifer to transmit water of the prevailing kinematic viscosity. The transmissivity, *T*, of an aquifer is equal to the hydraulic conductivity of the aquifer multiplied by the saturated thickness of the aquifer. Thus:

$$T = Kb \quad \text{(Eq A-4)}$$

Where:

T = transmissivity
K = hydraulic conductivity
b = aquifer thickness

Storage coefficient. The storage coefficient, *S*, is defined as the volume of water an aquifer releases from or stores per unit surface area of the aquifer per unit change in head. The storage coefficient is a dimensionless unit, in which the units in the numerator and the denominator cancel. The storage coefficient is given either in terms of a unit cube of the material or in terms of a unit prism of an aquifer, depending on the intended use.

The size of the storage coefficient depends on whether the aquifer is confined or unconfined. If the aquifer is confined, the water released from storage when the head declines comes from expansion of the water and from compression of the aquifer. Relative to a confined aquifer, the expansion of a given volume of water in response to a decline in pressure is very small.

REFERENCES

Heath, R.C. 1983. Basic Ground-Water Hydrology. Reston, Va.: USGS (US Geological Survey).

Abbreviations/Acronyms

ANSI	American National Standards Institute
AWWA	American Water Works Association
ID	Inside diameter
NGVD	National geodetic vertical datum (1929) – elevation to sea level in 1929
PVC	Polyvinyl chloride
SOP	Standard operating procedure
TDS	Total dissolved solids
USEPA	US Environmental Protection Agency
USGS	US Geological Survey

This page intentionally blank.

Index

Note: *f.* indicates figure; *t.* indicates table

H

I

L

M

N

O

P

Q

R

S

T

U

V

W

Z

This page intentionally blank.

AWWA Manuals

M1, *Principles of Water Rates, Fees, and Charges*, #30001
M2, *Instrumentation and Control*, #30002
M3, *Safety Management for Water Utilities*, #30003
M4, *Water Fluoridation Principles and Practices*, #30004
M5, *Water Utility Management*, #30005
M6, *Water Meters—Selection, Installation, Testing, and Maintenance*, #30006
M7, *Problem Organisms in Water: Identification and Treatment*, #30007
M9, *Concrete Pressure Pipe*, #30009
M11, *Steel Pipe—A Guide for Design and Installation*, #30011
M12, *Simplified Procedures for Water Examination*, #30012
M14, *Backflow Prevention and Cross-Connection Control: Recommended Practices*, #30014
M17, *Fire Hydrants: Installation, Field Testing, and Maintenance*, #30017
M19, *Emergency Planning for Water and Wastewater Utilities*, #30019
M20, *Water Chlorination/Chloramination Practices and Principles*, #30020
M21, *Groundwater*, #30021
M22, *Sizing Water Service Lines and Meters*, #30022
M23, *PVC Pipe—Design and Installation*, #30023
M24, *Planning for the Distribution of Reclaimed Water*, #30024
M25, *Flexible-Membrane Covers and Linings for Potable-Water Reservoirs*, #30025
M27, *External Corrosion Control for Infrastructure Sustainability*, #30027
M28, *Rehabilitation of Water Mains*, #30028
M29, *Water Utility Capital Financing*, #30029
M30, *Precoat Filtration*, #30030
M31, *Distribution System Requirements for Fire Protection*, #30031
M32, *Computer Modeling of Water Distribution Systems*, #30032
M33, *Flowmeters in Water Supply*, #30033
M36, *Water Audits and Loss Control Programs*, #30036
M37, *Operational Control of Coagulation and Filtration Processes*, #30037
M38, *Electrodialysis and Electrodialysis Reversal*, #30038
M41, *Ductile-Iron Pipe and Fittings*, #30041
M42, *Steel Water-Storage Tanks*, #30042
M44, *Distribution Valves: Selection, Installation, Field Testing, and Maintenance*, #30044
M45, *Fiberglass Pipe Design*, #30045
M46, *Reverse Osmosis and Nanofiltration*, #30046
M47, *Capital Project Delivery*, #30047
M48, *Waterborne Pathogens*, #30048
M49, *Quarter-Turn Valves: Head Loss, Torque, and Cavitation Analysis*, #30049
M50, *Water Resources Planning*, #30050
M51, *Air Valves: Air-Release, Air/Vacuum and Combination*, #30051
M52, *Water Conservation Programs—A Planning Manual*, #30052
M53, *Microfiltration and Ultrafiltration Membranes for Drinking Water*, #30053
M54, *Developing Rates for Small Systems*, #30054
M55, *PE Pipe—Design and Installation*, #30055
M56, *Nitrification Prevention and Control in Drinking Water*, #30056
M57, *Algae: Source to Treatment*, #30057
M58, *Internal Corrosion Control in Water Distribution Systems*, #30058
M60, *Drought Preparedness and Response*, #30060
M61, *Desalination of Seawater*, #30061
M62, *Membrane Applications for Water Reuse*, #30062
M63, *Aquifer Storage and Recovery*, #30063

- M65, *On-Site Generation of Hypochlorite*, #30065
- M66, *Cylinder and Vane Actuators and Controls—Design and Installation*, #30066
- M68, *Water Quality in Distribution Systems*, #30068
- M69, *Inland Desalination and Concentrate Management*, #30069
- M77, *Condition Assessment of Water Mains*, #30077

www.ingramcontent.com/pod-product-compliance
Ingram Content Group UK Ltd.
Pitfield, Milton Keynes, MK11 3LW, UK
UKHW052013041026
15590UKWH00008B/2303

9 781647 170400